Finding My Boundaries

Interesting People I've Met While Surveying

Eric B. Gladhill, PLS

Finding My Boundaries
Eric B. Gladhill, PLS

He may be reached at **gladhill.eric.16@gmail.com**

Cover design by Shane Gladhill
Cover photography by Lauren Gladhill

Interior photos provided by the author.

Book layout by Laura Ashton
laura@gitflorida.com

ISBN: 979-8510712940

Printed in the United States of America

This book is dedicated in memory of my son, Dusty, who followed in my footsteps for a brief period. You will never be forgotten.

Acknowledgments

I'm still trying to come to grips with the idea of being a writer. I don't feel like one and the term author was always reserved for the great writers that had their names on books in the library or the required reading list in high school, but here I am with a book that I wrote. I have many people to thank for this fact.

If my brother, Ralph, hadn't attended Penn State and acquired an associate's degree in land surveying, I wouldn't have known much about the profession. He also helped me with my trigonometry homework and made me realize why angles needed to be measured to the second (as in: degrees, minutes and seconds of arc). He was also the one who suggested that I try drafting after my art school failure; so thank you, Ralph.

Dave Davidson, current CFO of C.S. Davidson, Inc., (best boss ever—you'll find out later) told me that when he started his undergraduate studies, he considered being a writer instead of a civil engineer, like his father and grandfather had been. He reasoned that if he became a writer, he would never do any engineering as a hobby, but if he became an engineer, he could do some writing as a hobby. His featured article in the company newsletter was always a pleasure to read and so I knew that he could have been a successful writer, had he chosen that career. He also wrote some great letters and reports and through his editing of my letters and reports, I became a better writer, so thank you, Dave.

When I got my stories past 100 pages in a Word document, I felt like maybe I had something to publish. I was discussing my career one day with my dentist and told him about my book. He told me that his daughter had recently graduated from Shippensburg University with a degree in literature and she was looking for a job with a publisher. When I met Kelsey Kohler she was very happy to edit my manuscript.

This was the first professional editing of my book, so thank you, Kelsey.

My boss, Kerryn Fulton, current CEO of C.S. Davidson, Inc., was happy to hear that I had a finished manuscript and suggested that her husband, Wade, would take a look at it. Wade is a friend of mine who is a high school teacher of English Lit. Wade was willing to read through the pages and make some very useful suggestions on sections that needed re-writing. Any part of it that sounds good is due to his help, so thank you, Wade Fulton.

Thanks to the staff at the Pennsylvania Society of Land Surveyors and the Maryland Society of Surveyors (Lori Elliott, Khea Adams and the others behind the scene) for publishing some of my first excerpts from this book in the society's newsletters.

After getting some publicity in my surveying societies, I was confident enough to try submitting to the national surveying publications. Emell Derra Adolphus, Editor-in-Chief at *POB Magazine,* was kind enough to publish one of my excerpts and then ask for more, so thank you, Emell.

Likewise, I owe thanks to Scott Martin, columnist with *xyHt Magazine*, who offered me a guest column and thanks to Jeff Thoreson, Editor-in-Chief at that same publication, who published some more of my stories.

A successful surveyor-turned-author, named Patrick Naville, read one of my stories in *POB* and was kind enough to call me and leave a very nice voicemail, which complimented my writing and revealed that he had written some articles in the past for *POB*. When I called him back, I learned that he had also written some very good novels and had been nominated for the New York Times Bestseller List and had been named to the Amazon Best Seller List. As we continued corresponding, Pat told me about his self-publishing efforts and introduced me to the lady who would finally make this project happen. She gets her thanks next, but right now, I want to give a big shout-out to a surveyor who became an author. Please shop for his books on Amazon; especially if you were always disappointed in how the movie Butch Cassidy and the Sundance Kid, ended. He wrote an alternate ending sequel, entitled, *Echo Whispers*. Thank you, Pat.

The lady, who was introduced to me by Pat via an email, is Laura

Ashton. Without her help, this project would have foundered for some time. She was the one who pushed it over the goal line with lots of style and grace; answering stupid questions, sending information by email and formatting these pages for publication. Thank you, Laura.

I have to say thanks to all of the characters who are described in this book. Without the characters, what would a story be? Big thanks to the ones who gave me permission to use your full name. If there's a character that seems like it might be someone you know but with a different name, you're probably confusing them with someone else that you know.

Finally, I would have never completed this without the encouragement and support of my family. They have listened to my stories, read through some drafts and always got excited for me to get this book published. They've even encouraged me as I have begun my next book, a prequel about my early life. All my love and thanks go to Trudy, Shane and Keri.

Table of Contents

Acknowledgments 4

Foreword 9

Preface 11

Part I - Fox & Associates, Inc. 17

Part II - Associated Engineering Sciences, Inc. (A.E.S.I.) 55

Part III - Dewey Jordan 84

Part IV - Rettew 98

Part V - Fox & Associates, Inc. (second time) 112

Part VI - Buchart-Horn, Inc. 141

Part VII - Pinto Engineering, Inc. 156

Part VIII - Rettew (second time) 169

Part IX - C.S. Davidson, Inc. 190

Epilogue 237

About the Author 239

FOREWORD

by Patrick Naville

As a writer, I enjoy all different kinds of stories whether they be fiction or true accounts. If the story is well-written and it keeps me entertained, it's solid gold in my book!

When I read an article that Eric Gladhill recently had published in the professional surveying magazine, *POB*, it caught my attention for several reasons. It was not only entertaining, but as a land surveyor myself, it made me reflect on many different things I had encountered in my nearly 45 years working in that profession.

I contacted Eric to tell him how much I enjoyed his article and for two guys who never met before, we talked for quite a long time. We were like two old friends who hadn't seen each other in years, and it was remarkable how much we had in common, with one possible exception—I'm an "Old Fart" while Eric is still a relatively young man.

As we talked, Eric indicated he had a manuscript full of stories from his career as a land surveyor. He was interested in getting the stories published in book form but wasn't sure who to contact or how to go about the process.

When I told him I could hook him up with a professional who could help, I could hear the excitement in his voice. We exchanged email addresses and began communicating.

Eric was kind enough to send me a draft of his manuscript and asked if I would mind taking a spin through it just to get my thoughts on it. As I read it, I felt myself drawn into his stories. I felt like I was right there experiencing everything he was describing, and in a sense, I had been in similar situations at one time or another in my career.

Eric's stories not only captivated me, but I found myself laughing out loud at many of them! We both have had similar employers who were tyrants, and employers or co-workers who were wonderful mentors. The cast of characters Eric has worked with over the years and the things he's experienced would rival anything you would find on a popular TV sitcom. If I hadn't worked with a lot of strange birds myself over the years, I would have been hard-pressed to believe that everything Eric wrote was true, but trust me, folks—it is!

Eric's book, *Finding My Boundaries*, will not only be entertaining and enjoyable to all land surveyors, but also to everyone else who's worked in or around the engineering profession and construction trades—it's that entertaining and I highly recommend it!

Very well done, Mr. Gladhill—you are well on your way to a great second career as a writer!

Patrick Naville is a land surveyor licensed in five western states. He's also published seven books and is currently working on his eighth one. He was a New York Times Best Sellers List nominee for his first book, *Echo Whispers*, and made the Amazon Top 100 Best Sellers List for Westerns for his second book, *Cripple Creek*.

PREFACE

The profession of land surveying is rarely understood by anyone outside of the profession. Some attorneys, realtors, and developers have a basic understanding of the services that are provided by a surveyor, but the intricacies of the science usually remain unknown unless one has studied and practiced the craft. Sometimes an average property owner will need to have their property boundaries marked or perhaps they desire to create a smaller lot out of their land holdings (subdividing), which will require the services of a surveyor, but even then, they don't really understand everything that is involved with measuring land. What a surveyor actually does—in part—is to measure the surface of the earth and objects that lie on, above, or below the surface of the earth. Property boundaries are one of the most prevalent of those items that require surveying; in fact, a licensed surveyor, who has met all of the requirements of their state government, is the only person who can measure and determine the location of property boundaries. In this way, the surveyor has a quasi-judicial role in the legal aspect of property boundaries. There are also topographic surveys, which show the elevation contours or "relief" of the surface of the earth. These are mostly used by engineers and landscape architects to design proposed improvements. Location surveys are performed to create a basic map showing various features and improvements on the land.

The history of surveying goes back to the ancient "rope stretchers" of Egypt, whose math and methods were used to lay out the pyramids and other significant buildings, as well as property boundaries. During the time of Galileo, at the beginning of the 17th century, people began studying the universe through mathematics and surveying became more of a science related to astronomy. Two of the most famous surveyors in the United States, and well-known to everyone who lives near my native soil of Adams County, Pennsylvania, are Charles Mason and

Jeremiah Dixon. These men studied astronomy at the Royal Greenwich Observatory and traveled the world making measurements of the surface of the globe until being called upon to travel to the English Colonies of America to settle a land dispute between Lord Calvert (Maryland) and William Penn (Pennsylvania). The stone monuments that Mason and Dixon set between 1763 and 1768, still mark the legal boundary between these states. Many great books have been written just about this wondrous feat, which some historians have compared to the moon landing of that time, so I will not expound upon that history.

After this fantastic feat was accomplished, the colonies grew and as land grants were made to private citizens, surveys were made to "lay out" these tracts of land. One of the most famous surveyors in United States history was none other than the father of our country, George Washington. In 1749, at the young age of 17, Washington became the County Surveyor—a prestigious appointment in those days—for the frontier County of Culpepper, Virginia. His honesty and hard work gained him a lot of respect and notoriety among the citizens and leaders of government. He also earned a handsome wage doing this work, along with some land grants of his own. Some of this early experience certainly contributed to his leadership skills and popularity which created the career that we all know. Many other important surveyors helped to form our fledgling country in and around the time of its formation as an independent nation. Among those was another famous president: Thomas Jefferson. Abraham Lincoln, who is arguably the most influential president, who, through his actions during the Civil War, significantly changed the course of our nation's history, also served as a land surveyor. One of the amusing facts that many surveyors like to mention, about the grand sculpture on Mount Rushmore, is that the faces portrayed there are three surveyors and some other guy. This fact became a famous quote among surveyors due to the use of that phrase on posters, shirts, and other merchandise in professional trade magazines.

Surveying has always been an important profession necessary to protect the public and create logical and safe development of the land. During the post-World War II era in the US, land development and subdivision of land were being performed at an alarming rate. This led to some reckless surveying practices. During the age of the baby boom, people were able to afford the "American Dream" of owning a house, so building lots were being created by anyone with the means to speculate on some land. In many cases, the carpenters and builders, who had the measuring skills needed to layout and build a house, were also being employed to lay out the new property lines on the lots that were being sold. As one might be able to understand, the laying out of a single house of modest dimensions is different and much easier than measuring out whole blocks of lots with streets, etc. In my humble opinion, this led to the surveyor being regarded in the same manner as a builder, carpenter, or mason. All of this was written to give the reader some understanding of what the profession truly was and mainly, still is.

Many people that have some experience in construction believe that they have done some "surveying," when in fact they were usually performing some carpentry layout using engineering geometry. I think it is important for the general public to understand the profession of surveying because it truly is a mixture of mathematics, problem-solving, analytical thinking, research, and physical tasks that require a strong body. Surveyors have a lot of pride in their work and often exhibit the characteristics of a stubborn individual, mainly because their tasks and responsibilities are essential and mostly misunderstood.

It seems that surveying attracts some of the most fascinating characters. This is the main reason for this text; not that I have done anything of great significance in surveying, but the people I've met along the way have provided some stimulating and entertaining tales as well as leaving a lasting impression on me. Most of the people I've worked with were good, competent professionals; but for some reason, the position of rodman/chainman—the entry-level position on the survey crew—always attracted the lazy, happy-go-lucky types who were usually some kind of social misfit. I know the surveying profession has come a long way since the '70s, and I applaud the efforts of the surveying community, societies, and licensing boards to "clean up" the image of the surveyor and increase the public's appreciation of the profession. I still think we have a long way to go to get this point across and be fully recognized as true professionals. We must realize, too, that some of these "goofballs" eventually decided to grow up and become responsible licensed surveyors—I'm one of those!

The main reason that I wrote this book was to record the funny and remarkable stories that I've shared during my career of training and mentoring young surveyors and engineers. I would often repeat these stories as entertainment but also as valuable lessons learned. The young guys that I've worked with have told me that I am "geezin" (telling old stories like a geezer) when I am remembering the old days and talking about them. In fact, that's why I wrote this book, so that people would enjoy my stories and be amazed at all of the motivating events that I remembered. They even joked that I should write my stories down before I got too old, developed dementia, and forgot everything that happened. These stories aren't as exciting as some novels, but in many ways, they may be better because *truth is stranger than fiction.* I also thought it was a worthy method to preserve and archive my stories for my children and—someday—grandchildren.

I find it fascinating that considering my longevity in the profession (over 40 years), I've experienced quite an evolution in the equipment and practice of surveying. Mostly, for the old surveyors and those interested in the recent history of the practice of surveying, I'll mention some of the instruments used during my career. When I first started learning to operate the instruments in 1979, they were mostly steel Vernier transits made by Keueffel and Esser or W. and L. E. Gurley—affectionately called, "The Army Transit." The distance measurements were made with a 300-foot "drag chain," which was really a steel tape, but the "chain" nomenclature came from the ancient measuring tool for surveyors called a Gunter's chain. We had one electronic distance meter (EDM) at my first engagement in the profession and it was state-of-the-art equipment in the late '70s. I've reacted, retrained, and learned about the new equipment, such as total stations, data collectors, and the various GPS equipment. I won't make this a story of the evolution of surveying equipment. Someone else can do that; this is about people. I hope that other surveyors will enjoy these stories because they can relate to the experiences, but I really hope that this text transcends the experiences of surveyors and also relates to the average person on a humanistic level. As you read through this, I hope you enjoy the experiences as I have! I hope you can see the faces and hear the voices as I often do when remembering these incidences.

Part I

Fox & Associates, Inc
January 1979 – July 1981

My first job in surveying came by accident or perhaps part of a grand plan from God that I didn't expect. I was only 17 years old and I had dropped out of art school after just one semester of failing grades. I had tried art school simply because it seemed easier than studying math in college, and I didn't have to take SATs to be accepted at the Maryland Institute College of Art. The only course that I had passed at that school located in Baltimore, Maryland, was lettering. In this course, we were given explicit instructions on how to create certain styles of printed letters of the alphabet.

Before the advent of computerized word processors and various design programs, such as AutoCAD, artists were relied upon to create graphic designs which included lettering to spell out a message. Imagine the old posters or printed advertisements with various styles of fonts. These were being created by an artist. People didn't have a pull-down menu of fifty or more fonts to use to create a message on a computer program or application—*app*, as they are currently called. I didn't realize it then, but this course appealed to me because it was structured with rules of dimensions and styles.

I had done well in my high school woodshop class when we drew our plans for the projects that we built, so I guess I had some proclivity for this type of drawing and labeling used in drafting. My brother Ralph, who was already a Penn State graduate with an associate's degree in surveying, encouraged me to apply as an entry-level draftsman (old, gender-specific term of that time) with several of the engineering/surveying firms in Hagerstown, Maryland. He was also the one who had the drafting tools and helped when I was designing my mahogany coffee table for woodshop class. Ralph was employed at an engineering firm in Hagerstown at the time and he knew some of the firms that might need drafting help. During that era, as stated previously, there were no computer programs or apps, which allowed a designer to create drawings for construction or manufacture of proposed improvements or products; the designer relied on a draftsman to draw the lines and dimensions, at a prescribed scale, of the element being designed.

I stopped in at Frick Co. in Waynesboro, Pennsylvania, first just to see if they were interested in hiring a bright, young, talented draftsman with no experience. They took my application and that was the end of that.

My second stop was at the offices of Fox & Associates, Inc., located then at 405 N. Potomac Street in Hagerstown, Maryland. After filling out an application for the position of draftsman, I was interviewed by Bernie Charles, who was part-owner of the firm. It was never clear who was in control at Fox & Associates because the firm bore the name of William E. Fox, the apparent owner, and impetus behind the company's day-to-day activities. It was said that Bernie Charles, who was a licensed surveyor in several states, had inherited the money necessary to run a business, and Bill Fox, who was also licensed, had the business sense and savvy to run the show.

When Bernie interviewed me that day in mid-January, all he could say about my experience—which only included high school, one or two short-lived summer jobs, and one semester of art school—was, "Another Pennsylvania boy, huh?" He then asked if I was using any drugs and I nearly fell out of my chair. This was 1979 and he told me he always worried about these young boys that they were hiring. I simply and flatly said, "No." The truth was the exact opposite. I had a lot of experience with drug use at the time. The whole truth was that I had failed in art school because of excessive drug use. After meeting some of the guys that were working there and hearing stories about some of the guys who had worked there in the past, I understood his concern!

After about ten minutes of small talk, Bernie finally asked me if I would be interested in working in the field on a survey crew. I couldn't believe it. I was sure that a person would have to go to college to work on a survey crew. I told him I loved to work outdoors, though I don't know why I did. When I think about it now I always remember the cold, windy days when I stood behind an instrument squinting tears out of my eyes to see through the scope, and the hot, sweltering days spent cutting brush, pounding hubs (a wooden stake, that is driven flush to the ground where a tack can be set for a precise point) or climbing a mountain with lots of equipment on my back to get to a distant property boundary. But back then I guess working outdoors meant spending time with my dad cutting firewood or tilling the garden. Outdoors also meant playing baseball, hunting or fishing, so this offer that Mr. Charles was making me sounded great! Then he warned me that we would be working in Moorefield, West Virginia, about four hours south of Hagerstown. I was informed that the motel room would be paid for and I would receive a per-diem. I think it was about $18 back then. I can't imagine living on that since it's not too hard to spend that on one good dinner today.

I was excited and intrigued to be afforded the opportunity to start work in a prestigious profession like surveying. Without knowing the full range and depth of knowledge that was necessary to be a surveyor, I already knew enough from my conversations with my brother, Ralph, that this profession required a lot of knowledge of trigonometry and it was an important undertaking. I had seen a Mason and Dixon milestone, near to my home, which had been set on the Maryland and Pennsylvania state line to settle a land dispute. I felt like I was going to get an opportunity to work at a job with some dignity and pay off those student loans.

Entry-level job

I agreed to the job even though it paid minimum wage—I think it was $3.35/hour, a meager salary, even back then. I met with Bill Fox later that week and he told me how important being a rodman/chainman was. The job we were going to work on was a federally funded, residential improvement project for a depressed little backwoods town, sort of like where I went to school—the quaint borough of Fairfield, Pennsylvania. The surveying would be preliminary engineering surveys, such as cross-sections and baseline right-angle location. Bill even gave me a textbook to study. This was serious business! I'm still thankful for that advice and that first job, though. It got me started on the right path with some appreciation for the mathematics involved. I took that book home and read about the applications of some of the trigonometry I had learned in high school. I went to my brother Ralph's house and asked him some questions about surveying. He explained what I would be doing in the field and how the data (distance and elevations) would be plotted by a draftsman to create a scale drawing of the actual cross-section of the earth's surface. I was excited to begin this journey.

Monday morning, January 21, 1979, my parents took me to Waynesboro, where they were employed in a machine shop, to meet Benny, the party chief, with whom I would be working. The job title of *Party Chief* has always caused some smiles and giggles to the reference of "party." People would joke about the title and maybe even consider that the party chief was the guy who bought all of the beer, pretzels, and cake, and may even send out the invitations. The old term for the survey crew was "survey party." There were even some warning signs to be used along the highway when the personnel were working along the shoulder of the road that had the words, "Survey Party Ahead." This usually elicited a few comments from passing motorists about the location of the party or asking where the beer was. Some companies eliminate the confusion and amusement by terming the position of the leader of the surveying group in the field as crew chief.

Benny lived near Waynesboro and agreed to pick me up on his way into the office. Benny was a graduate of Penn State's Mont Alto campus. Penn State has a good program for surveying studies but it no longer offers that program at the Mont Alto campus. When the surveying program was held at that campus, there was always a good supply of survey interns employed at the local Hagerstown firms. We used to call them "Mont Alto boys." For the most part, they were smart kids with an attitude about us guys enrolled in the "school of hard knocks." It was a point of constant contention between both groups who were proud of their career paths. We, "the uneducated" felt we were superior because of our real-life experience, and they, "the book-learned" felt they were the best because they had learned some calculus and read how to determine true north from taking sights on Polaris ("celestial observations" they called it). The funniest thing that we used to pick on them about was the "Mont Alto Sight." One of the main objectives of showing the instrument operator where a point is located, by holding a plumb bob directly over the point (commonly referred to as "giving a sight" or "giving line"), is to hold the string steady. Of course, it is common logic that the shorter the string is, the steadier the hold will be. The Mont Alto students consistently insisted on holding the string of the plumb bob at least as high as their head. The result was very shaky sights.

Benny had worked in the field for four or five years at the time so he was well-rounded with education and experience. At the office in Hagerstown, we met the instrument man, Dudley, and the other crew who would be doing the topography. There was Mike, the young party chief-in-training and his equally young and inexperienced instrument man, Kent Faughander. The rodman was a draftsman-turned-rodman that we called Radar because he resembled the character on the then-popular TV show *M*A*S*H.* Bill Fox and one of the engineering designers went along down with us to the job site to look around and see the project.

Dudley was quite the character. I should devote a whole chapter just to his personality, but I will describe him in several anecdotes. He was from Hedgesville, West Virginia, so he was somewhat at home in Moorefield even though it was hours from his boyhood home. The first day on the job, Dudley and I were busying ourselves with putting tabs of flagging on nails to be set for the "baselines." He began to tell me how Benny could be a difficult boss to work for. Apparently, he and Benny didn't see eye to eye. Dudley was more interested in having fun and relaxing than getting the job done right. He pointed out the fact that Benny had no eyelashes! "Really," he said, "check it out when he comes back, there's no eyelashes." I could tell this was going to be a real lively crew. Dudley told me if Benny ever gave me "a bunch of shit," I should just tell him to "go to hell, you no-eyelash mother****." I didn't think that would be necessary but I filed it away for future reference.

Oddly enough, it didn't take long to bring it into use. That afternoon as we were running a level loop to establish our datum elevations (which was National Geodetic Vertical Datum of 1929 – NGVD 29) on the project. This was supposed to be some accurate elevations with benchmarks being established so we used an automatic level and the old “Frisco rod,” a three-section wooden rod. I was running the rod and Benny came walking by me and said, “Is that a banana you’re eating, or is that your nose?” At that point in my life, I had no qualms over using colorful language that Dudley had instructed me to use, so I repeated the words as I had been taught. Benny’s jaw dropped open and he couldn’t think of anything to say! I guess it wasn’t the best way to get started with your boss, but I was not quite eighteen years old, and I didn’t have much respect for authority then.

Working in Moorefield, West Virginia, in 1979 was a bit of an adventure. There were no fast food places to grab a quick lunch, so we found the two family restaurants in town where we could eat. At lunchtime, Benny looked at me and asked, “Do you like to drink?” I answered in the affirmative, even though I was not going to be eighteen years old for another four days. He then asked, “What do you like to drink?” I answered, “Anything alcoholic.” He made a wry grin while Dudley was laughing hardily. Then he asked, “How about narcotic?” I feigned ignorance and said, “I don’t know what that means.”

After work and before we went to the motel, we went to the ABC store, which is what they called the state liquor store. Benny asked me to pick something out and he would buy it for me. For some unknown reason—I guess it was my ignorance of what was considered good liquor—I pointed to a small bottle of lime vodka. He told me to go out to the truck. When I reimbursed him for the purchase, it was about $3. Everyone else was getting Jack Daniels or some other hard liquor because the beer was 3-2 beer, meaning that the alcohol content was 3.2%. That was what was allowed for purchase by anyone over eighteen years of age.

We sat around the motel room drinking our alcohol and playing some cards. After a while Mike said, “We’re going for a ride.” Benny said that he could, “smoke some of that stuff” in the room, he didn’t care. Mike told him that wasn’t the reason, he just wanted to go for a ride, and then he pointed to some of us and asked if we wanted to join him. I was chosen and so I went along. Soon after we got in the truck, a joint was produced from a pocket and we started smoking it while driving around the back roads of Moorefield. When I asked Mike why he didn’t want to smoke in front of Benny, he said that he didn’t trust him and thought he would use that information against him later. They all told me not to trust Benny.

We worked four, ten-hour days and we were going to drive home on Thursday night, which was my eighteenth birthday. On Wednesday evening the guys decided to visit a private club that they had learned about. They were told that the only establishment that could serve alcohol in that region were private clubs where membership was required. Since I was not legal age, I stayed at the motel with one of the young guys who didn't drink.

The next week, they offered to take me along to this club named "The Caledonia Club." The membership cost $1 and you had to write your name in a "membership book." Now I felt like one of the men on the crew as we played billiards (shot pool) and drank copious amounts of beer and liquor. The management didn't seem to care if I was twenty-one years-old or not. I wasn't very good at shooting pool and Dudley noticed. He comforted my ego by saying, "You're a lot like me, we didn't grow up with a game room in the house; we were always out in the woods shooting rabbits and stuff."

We did a lot of work during the day, but we always drank and had fun at night. This was a bit of a comfort to me; I was just 18 and away from my family and friends. If I were at home I probably would have been out with my friends who wanted to drink beer anyhow, so this was a good compromise to be out with the "work buddies."

While we were working in the "New Addition" of Moorefield, (this is what they called the area where we were working) we saw some interesting dwellings. There were some old homes, some new homes, and several mobile homes. Our project involved federal funding to upgrade the infrastructure of this area. It really needed it.

Apparently, the streets were laid out without any real design. The whole area didn't have any good drainage and some of the yards flooded with a bit of rain. There was one house where we observed a storm door open and a kid on a Big Wheel trike rode out of the house and down a ramp. Soon the door opened again, and a kid goat came out of the house!

Mike's crew was doing the right-angle baseline location and each feature had to be located with distances from the baseline and labeled in the field notes. There were some odd dwellings in that neighborhood, and they needed to be noted with their type of construction. It was normal to see labels such as, "Two-Story Frame Dwelling" or "One-Story Brick Rancher." Mike looked at one of the wooden frame building that was constructed of weathered boards, with no basement and several window frames covered with plastic sheets. He labeled one as a "One-story Hillbilly Shack."

Weeks later, he was lectured about how he couldn't make judgements on people's dwellings. The drafter had written the description verbatim from the field notes onto the base mapping sheets of plans. Bill Fox was outraged at the thought that they may have met with the town council and rolled out the plans to reveal that type of description—the owner may have even been on the town council. This was corrected but we all got a message about how to show respect for these good folks of Moorefield.

The local paper covered our surveying work that was being done. That newspaper even had some posed pictures of us doing our work on the front page. I still have a copy of that issue of the *Moorefield Examiner*. This was big news! Especially when one considers the headline on that issue, "Heishmans Have Another Girl."

Drill Sergeant Party Chief

As I got to work more with Benny, I realized that he could be a tough boss. He was very much no-nonsense and tried to be the "big man" in many ways. Apparently, he had been in the Army Reserve or something and did his best to be the commanding officer of the crew. His stand-by phrase was "rank has its privileges." He usually walked around with his tool pouch on and didn't carry much else. If it was summertime, he would hold the rear end of the chain and let someone else head-chain and do all of the stake pounding. If it were winter and it was cold out, he would do the head-chaining so he could hammer the stakes and stay warm.

I was happy when I worked with one of the other party chiefs because they were fair and taught me a lot, but Benny always kept secrets about the job and just barked orders. About the time one of us would be ready to lash out at him, he would start to act like your friend and see if he could smooth things over. He enjoyed a card game called "F*** your Buddy." I always thought that suited his personality!

I will admit that Benny could be a nice guy away from work. When we showed up for work and it was raining, we were guaranteed two hours of pay for "show-up" time. This meant that we would normally do maintenance on the tools and clean out the truck. It was easy work and we would smoke cigarettes in the garage while we cleaned and oiled the tape measures and replaced tool handles while we listened to some rock music on the truck radio. We also sharpened cutting tools, such as machetes, brush hooks and axes. One time on a "rain-day" Benny offered that we could go to his house to clean the truck and do repairs to the tool box. He said that he would set up a full bar in his garage and we could pop tapes into his home stereo. We had a lot of fun and enjoyed the hospitality. Unfortunately, this was all gone when we had to work together again.

One good example of how Benny acted when we worked in the field was on a big industrial park project that was just getting started. We had to set aerial photo control to have topographic mapping produced. When we arrived, Benny never told us what this job entailed; he just got out of the truck, got his tool pouch on, and started barking orders to Tim (the rodman) and me: "Get two stake bags packed with hubs, guard stakes, panel material, sledge hammers, bull pins, and brush hooks." By the time we got all of that together, Benny was gone!

I thought I had seen him going down over the bank beyond the guiderail along the road so I called out for him. He answered and we went hurriedly toward the sound of his voice. When we got about 200 feet into the woods, I called for him again. No answer. I called out louder—about as hard as I could yell—and heard no reply. We walked quickly into the woods and yelled his name. When we heard nothing, I sat down on a downed tree trunk and motioned Tim to do the same. We sat there and waited. We exchanged our criticisms of how a boss could be such a prick! Eventually, we heard his voice calling out to us. I indicated to Tim, with a finger against my lips, that he should stay quiet. I told Tim that if he wasn't going to answer me, I wasn't going to answer him. We waited until he was just about to walk by us; about 100 feet through the brush, until we said, "Hey!" He walked over to us and asked us what was going on: "Why didn't you answer me when I yelled?" I told him that if he was going to walk away from us and not answer us, then that is what he deserved. He was not happy.

Shortly after that episode, some of us were discussing this same subject over a few beers—probably on a rain-day—and we decided that we needed to have a confrontation and explain how we felt. Someone with a voice of reason suggested that this was our only hope to improve the situation so we arranged a meeting one day after work hours. Mike led the discussion and then we all "piled on" with our gripes. Benny looked like a whipped pup and apologized. This was a good lesson for me, as I began to see that most times when a person is acting aggressive, they don't even realize that they are being an aggressor or doing anything that would cause duress to others. This "intervention" did not have any long-lasting effects.

At one point during a slow period, we were working 32-hour weeks with Mondays being our day off. I had acquired a bear tag along with my hunting license that year. In those days the bear season in Pennsylvania was one day; the Monday before Thanksgiving. I had planned to travel with my Dad and brothers to stay in a motel in Dubois, Pennsylvania, on Sunday night before the season to hunt the next day. The Friday morning before our planned excursion into northern Pennsylvania, Benny told us that we were going to start working five days a week, starting with the coming Monday. I advised him that I would not be able to come to work because of my hunting trip. He told me that I was expected to work when work was available. When I protested, he told me they had a lot of applications for work on file and that they could find someone to replace me. I countered with, "Go ahead; you'll just make me get a real job!" We had to work together in the field that day and after a few hours of carrying the transit up and down through the woods, Benny asked me if I wanted his help by offering to carry the transit. I asked him why he wanted to be nice to me now. He said, "Well, I was pretty mean to you earlier, I just wanted to make up for it." I told him, "No thanks, you should just try being nice to people, then you don't have to make up for it." Such was our relationship.

The West Virginia Hillbilly

Dudley was one guy who always surprised me. Once when we were cutting a centerline for a new county road, we encountered lots of multi-flora roses. Dudley decided to make short work of the cutting by bringing in his chain saw. Normally, Dudley stayed behind the instrument to show the "cutters" where the line was located but when we got in thick briars, he leaped forward, pulled the cord to start the chainsaw, and yelled, "Let Beowulf through!" We all jerked in surprise! As the chainsaw ripped into full throttle, he pulled his pullover hood completely over his head and just waded into the thick, sharp thorny bushes. He usually came out with briars stuck in his pullover, and blood trickling from his arms. Of course, he was always grinning like a maniac, so that just made his appearance eerier.

One time he displayed some other maniacal behavior while walking along a street in an industrial park. As the party chief and I walked ahead of Dudley we heard him let out a low growl. By the time we turned around, he had squared off with a telephone pole using the brushhook that he was wielding to hack at the sides of it. He told us it "looked at him wrong!" This was typical Dudley to us. We never knew when he would do something strange.

His strange behavior started in his youth. He told us his son's teacher had requested a parent-teacher conference because the son had been going into the bathroom at school, pulling his coat over his head, and making animal noises. The teacher said he was scaring the other children. Dudley said, "Oh, heck! I used to do the same thing." The other guy on the crew and I exchanged knowing glances as we realized that Dudley was different than most people and considered it normal. He was a nice man, so we just enjoyed his "different" behavior as amusing and entertaining.

The Mentor

One of the party chiefs at Fox and Associates who was nice to work for was Bob Banzhoff. He liked to refer to himself as Trebor Fohsnab (Robert Banzhoff spelled backward). I never imagined that him telling me this fact would lead to my fascination—almost compulsion—to look at words backward. I really enjoy odd combinations that actually make sense backward; like Wal Mart becomes Tram Law. Okay, I'm a bit weird, but Bob started it!

Bob was always teaching. He was used to being assigned,

inexperienced, dead-end kids who might comprehend and retain the knowledge that he was imparting, or they may just go work at a car wash next. When I started working with Bob, he would show me how to reduce slope distance to horizontal distance and read the vertical angle on the vernier. When we were doing a construction stakeout at the Woods Resort in West Virginia, he explained the principles of slope staking. It was like a field university!

Bob Banzhoff Taking Field Notes

We had a large, ongoing project in La Vale, Maryland, where we were charged with performing as-built surveys on the Country Club Mall. This meant that we had to spend the night in the nearest motel during the weeks that we worked there. Of course, my thoughts at the time were, "Oh boy, out of town!" I didn't have to report back home to my parents' house, so there were no rules after work. We received about $15 per diem for meals. I would usually spend most of it on beer.

We would always start out early on a Monday and stop at a little truck stop along US Route 40 on Polish Mountain. This was in the days before Interstate 68 so it was a scenic trip over the mountains and Bob would regale us with tales of his country band in which he played drums. I was a frustrated, under-practiced, and untalented drummer at the time, so I always wanted to hear what it was like to play professional gigs in bars, even if it was country—definitely not my style of music.

It was during those nights staying out of town at the Toll House Motel—don't look, it's not there any longer, but the Toll Gate House is still there and marked with a sign placed by the Maryland Historic Trust—Bob would teach us how to compute deflection angles and chord distances for staking out curves, while I was swilling a Mickey Wide Mouth or Elephant Malt Ale. He had patience with us and even taught us some right triangle solutions and basic coordinate geometry! It all made sense to me because I did well at trigonometry in high school.

When we stayed overnight in LaVale at the Toll House Motel, we usually had a few beers before dinner and then ate at a local restaurant. As I remember it, there weren't that many restaurants that served beer. There was either the bar where you could drink or the mom-and-pop restaurant where they served coke and other soft drinks. Of course, we always had a fast-food place here and there, but this one restaurant in that town was called D'Atri's.

This place seemed like a Greek-owned diner, but it had a definite Italian flair. They served pepperoni rolls and pepperoni omelets. As with a typical Greek-owned diner, they had the full bakery with everything from pastries to donuts to full-sized wedding cakes. This place wasn't just great for breakfast; we sometimes went there for lunch and dinner. One of the best items on the menu was their cheesesteak sub. Of course, they used their baked-fresh-in-the-store rolls and good steak, but the specially seasoned mayonnaise and their oil and vinegar dressing on top of a full salad's-worth of lettuce and tomatoes on top just made it so delicious! The amount of lettuce was somewhat overdone, but you could knock a little bit of that off and still have a nice side salad with your sub.

I can honestly say that my mouth is watering as I write this, just from the memory of those subs. I'll admit, it was not that long since I've tasted one because the restaurant is still there on Scenic Route 40 in LaVale and I've stopped by several times over the past 20 years. It now has a second floor to the place and I love taking family and co-workers there to regale them with my stories of eating there in the "old days."

In the evenings, after we had dinner—or maybe just skipped dinner if we were going to drink plenty—we would sometimes go to the Toll House Inn Lounge. This was one of those motor court bars that was a handy place for the traveler to have a few drinks close to "home." I guess it was a holdover from the days of the old inns on the stagecoach routes where a traveler could get a meal, drinks, and bed for the night; all in one stop.

This was just down the hill from Frostburg State University and so the establishment would have bands, even on weekdays. If there was a band playing, Dudley and I would go there and have a few drinks. We might dance with a few of the college girls from Frostburg State, but mostly we were just killing time and having fun.

One night they had a drink special on banana sombreros. Dudley loved a bargain and he started drinking these concoctions that I could only assume contained tequila and banana liqueur. After about three or four of them, he started dancing with the girls, which provided plenty of entertainment! He had a few more before the night was over and I had to guide him a bit to get him back to the room.

The next day was a rough one for Dudley; not only was his head hurting but he had to take continuous breaks into the bushes as we were doing a boundary survey for part of the Country Club Mall. Bob Banzhoff and I had a lot of fun at his expense as we made jokes about him running off in the bushes with the TP (a term used interchangeably for turning point—a point in a level loop—and toilet paper) every thirty minutes or so. I kept asking if he wanted to go drink some more banana sombreros. I don't think he drank anything that night!

Sometimes when Bob was out late, either playing drums or hanging out with his new girlfriend in West Virginia, he would be semi-asleep during the day. We used to talk him into taking a break in the middle of the morning. We would all pack two sandwiches in our lunch and we started calling our breaks "the ten o'clock Sammy (short for sammich—actually sandwich) break." Bob thought it was amusing sometimes, but other times he would get pretty pissed off at us for starting to mention the ten o'clock sammy break at 8:00 a.m.

Some of those days when he was really tired, Bob would drift off in the middle of a sentence as we were sitting in the truck for our mid-morning break. It was so funny because he was the only one talking and slowly, the words would slow and then stop, his head would droop, and then snoring would start. The instrument man and I would start to snicker and then Bob would slowly regain some consciousness and sometimes slowly raise *one* eyelid and look at us. Then we would start to laugh uncontrollably!

Bob told me years later that after he continued to struggle with feeling sleep-deprived, he was diagnosed with sleep apnea so even when he got to bed at a decent hour, he was not getting the rest that he needed.

One of the ongoing projects which I worked on with Bob Banzhoff was The Woods, near Hedgesville, West Virginia. Bob seemed to really love this project for several reasons. First, I think he liked the drive to the job site. We'd go down Interstate 81 to Martinsburg and then go west on Route 9. The whole trip took about an hour. Bob always wanted to hurry at our daily coffee stop in the morning because he didn't want to miss "the red bug lady." I didn't have a clue what that meant on the first trip. The other guy on the crew, Eric Kersting, just laughed (he rarely talked). As we drove down the two-lane road and passed a school bus, Bob sat up in his seat and said, "Here she comes." As this red Volkswagen Beetle drove by in the other direction, Bob leaned forward and waved. The cute young lady who was driving the VW, waved back and Bob got a big smile on his face. He was having a relationship with this woman whom he'd never really met, but they exchanged a smile and a wave every time that we went that way!

The Woods was a vacation home development in—well, as you may guess—a forested area. There were hundreds of acres of wooded hills that were being cleared for roads. The developers wanted to save as many trees as they could, or the name of the development wouldn't make much sense. To that end, they wanted us to stake the road centerline and then set slope stakes for the construction of the road. Slope staking—affectionately termed *slop* staking by me—is a method of obtaining the elevation of the existing ground at the "hinge point," which is either the top of slope (in a fill section, where the proposed elevation is above the existing elevation) or the ditch line in a cut section (where the proposed elevation is below the existing elevation). It was good training for me and I soon caught on to doing the math in my head and determining if we needed to move the rod in toward the centerline or out away from the centerline.

The difficult part was carrying the equipment and stakes into the remote areas of the hilly woods where the roads were being proposed. The interesting thing that was new for all of us was the use of a magnetic locater to find the old survey control points in the woods. The original surveyor of this land was a partner in the development. He would provide his field notes and a Schonstedt-Heliflux magnetic locator. Every survey crew has something similar these days, but back then—and this seems unimaginable—we didn't have that technology and we spent hours digging around in the dirt for metallic survey points. In order to save some labor and time in carrying the stakes in to the work area for the day, we always wanted Bob to drive the old GMC Suburban in as close as possible.

There was one section of the development that required us to drive across a major drainage swale. I guess the developers hadn't yet placed a culvert that would eventually be needed, so we just had to ford this small stream when we were experiencing a rainy season. After several strong thunderstorms, this drainage feature became more of a swamp than a drainage; it sat stagnate and smelled like decaying vegetation—imagine rotting celery or potatoes .... The first time we had to venture to the area beyond the swampy bog in the middle of the road, Bob felt it was best to park the truck and walk around it. This meant carrying armloads of stakes along with the magnetic locator, instrument, measuring tape, rod, hammer, machete, etc. After a full day with several breaks to hike the quarter-mile back to the truck to get more supplies, we realized we had lost a lot of time.

The next day, we convinced Bob to drive through the "swamp." It was about twenty to thirty feet across and maybe six inches of water over a mucky base that went deep (we had no idea how deep yet). We sat in the truck at the top of the hill on the old roadbed which sloped down toward the sloppy mess at the bottom. Bob expressed his doubts and concerns and even mentioned the real possibility of getting the truck stuck. I said, "Heck Bob, we get enough speed, we'll just skip right across this mucky water." He revved the engine a bit dropped the floor shifter into first gear (first was down, but "L," which we called Granny Gear, was up—only used for steep hills) and accelerated. We probably hit thirty-five MPH as we splashed into this horrid, mucky water. The truck skidded slightly right and then left as Bob corrected the steering into the skid. We made it through, but the whole truck was covered with light brown, slimy goop. It stank to high heaven. Again, I remind you, imagine rotting vegetables. We were disgusted but happy that we didn't have to carry all of our equipment and supplies. We had to do the same to return at the end of the day and Bob didn't go as fast. Perhaps he was still a bit shaken up from the wild ride in the morning, but we still made it through.

The next day when we returned it was a bit different. I'm still not sure why, but Bob approached the swamp area slowly. We gave some mild encouragement to speed it up, but I guess we didn't get too excited because we trusted Bob's judgment. As we started into the edge of the water, he slowed even more and then started to accelerate. We immediately stopped all forward progression and the rear tires dug ruts and we sank. Not only was the axle resting on the ground but the whole frame was sitting on top of the mud! This was not the first time that I had seen a Suburban stuck in the mud, and I knew this one was worse than the others that I had witnessed.

These were two-wheel-drive trucks that we used back then. They also had about a half-ton of equipment and supplies. We used 1-inch pipes for property corners and we filled the bin in the toolbox every time that we had down time at the office. When these rear-heavy trucks got into the mud and the tires started to spin, it would just dig ruts and sink further in as a result of the weight. We used to say, "These things will dig their own grave."

As we stepped out of the truck we sunk up to our knees in muck. Now we knew how deep it was. I said, "I'll go get the bulldozer operator to bring his machine down to pull us out," Bob said, "No, we can dig this out!" I was incredulous. You could just look at this truck with the bottom of the door frame resting on mud and Bob says we can get it out ... *"How?"* I asked. Bob said, "Get the shovel and some stakes; we'll dig trenches back on a slope to dry land and lay stakes in them to drive on—we'll get it out." Imagine my dismay at doing extra work that I was not counting on doing that morning. We dug and we cursed and we dug some more. If you've ever dug sloppy muck, you realize this is not easy. After fifteen minutes with little progress, I said, "Okay, now can I go get the bulldozer?" Bob said, "No." I think it was a matter of pride; he didn't want everyone on the job site to know that he had failed. We worked for an hour and managed to completely bury about 40 stakes into the mud. As soon as we tried to move the truck onto them, the ends sunk into the mud, and the tires, which had their treads filled with mud, started to spin. Of course, it was fun to watch the smoke show as Bob revved the engine and burned rubber on the wooden stakes!

Finally, when I asked for about the fourth time, "Now, can I go for the bulldozer?," Bob, said, "No, I'll go." Off he trod and we sat down under a tree for an extended smoke break. We didn't try crossing that spot again until it was completely dried. Bob was usually a great driver but, he never would fully answer my inquiries about why he approached that spot in that way. He just said he was concerned about getting stuck!

Davey, King of Thorazine

There was a new guy who started working on the crew one day with Mike, Dudley, and me. His name was Dave Thomas—the same name as the guy who started Wendy's fast-food restaurants and named it after his daughter. This Dave was quite different. While most of us in the survey department were around twenty to twenty-five years old, Dudley was about thirty and we thought he was old! Well, Dave was about thirty-two and he acted forty-five. He was slow—both in his gait, his actions, and his wit. He had no sense of humor and actually looked like a beat-down, slow-witted, George Costanza, the character from Seinfeld. When he came to work each morning, he had on the same denim carpenter's pants.

One day Dudley came in and said, in his most exuberant voice, "Hey, Davey, are you still wearing that same pair of pants?" Of course, we all laughed at the new guy getting harassed.

I soon found out that he lived in Littlestown and my parents' house was practically on his way. Someone (probably Benny) suggested that we ride together and before I could say, "Uh, no thanks," I was sharing a ride to work with this weirdo.

I eventually learned from him that he used to do drugs, but I couldn't tell if he was joking or not because he had no emotions at all.

We talked about tripping on acid and he displayed his wry grin, but not the laughter that I would usually hear from guys who had tripped before. I eventually asked him what drug was influencing his behavior and he told me it was Thorazine.

I started to research how that drug was used—and back then, there was no internet. I recently Googled "lobotomy in a pill" and found the name: Thorazine. I also found the other nickname for this drug: "liquid straight jacket." This guy's actions were exactly what I just read about the drug. He had no emotions and talked in a detached and eerie voice. He never knew what joke we were laughing at, but when we would look over at him during a lull in the conversation, he would have a little wry smile on his face. If I asked what was funny, he would answer in slow, measured meter, "What do you mean?" while still smiling to himself.

When we left him standing at a back-sight point, we would focus the instrument scope on him and see him smiling widely, as he puffed on a cigar. We would holler and wave a traffic flag to get his attention for a sight, but he would just look at us and smile. I slowly realized that he really was on drugs; legal prescription drugs that the doctors gave him to make him somewhat normal.

Once when I was riding home from work with him—since it was his turn to drive for the week—he was driving his old '64 Chevy Chevelle, the traffic was pretty heavy and he was breathing hard with stress, sort of hissing air out through his teeth. I asked, "What's wrong?" He simply mumbled something like, "I can't stand this," in a terse way. After a while, I saw him gripping the edge of his seat so hard that I thought his fingernails would rip right through the vinyl and pull out a chunk of the foam padding! I asked if he was alright and he just spat out the word, "Traffic." I thought it was best to just keep quiet and not aggravate him anymore. I also decided it was best to not ride with him in the future.

I did a poor job of lying about why I no longer wanted to share a ride and he questioned me more than necessary, but eventually, he pulled away from my home and I never rode with him again. He only worked there for another month or so. They confronted him about his abilities to do his job while under the effects of his prescription drugs. He decided that he should just quit and I never saw him again.

The Old Professor

Another character and quite an experienced surveyor who worked for Fox & Associates was Ralph Donnelly. Ralph was an old man back then! He passed away in 2003 and has been honored in various ways, such as the Maryland Society of Surveyors named a calibration baseline after him. The year was 1980 when he managed the Martinsburg, West Virginia, office of Fox & Associates. The crew that I worked with, led by Mike, was assigned to the Martinsburg office. We had some interesting adventures working for old Ralph.

I never knew all of the details, but I was told by various people that Ralph Donnelly was a pioneer in writing survey programs for hand-held calculators. These were more like hand-held computers, produced by Texas Instruments and named the TI-58C and TI-59C. They had a magnetic card reader that could be used to write programs (actually just record keystrokes, with pauses for the input of the variable data) that would compute coordinates or curve geometry, to name a few. Supposedly, Texas Instruments paid Ralph for some of his programs. This was a man that was born in 1910 and had figured out and mastered the latest in technology for field survey computing! I was very impressed by his experience and knowledge, but I found out soon that Ralph had some quirks.

Once when we met him at a job site, he showed us a fenceline and told us to survey that property. We asked for a deed or deed plotting or something—maybe even a sketch of the property. Ralph said this property lay between the fence and the railroad that we had just crossed on the way back the road. We asked how deep the property was and Ralph said, "Just go back along the fenceline until you get tired and take a right." We laughed and made some remarks about getting tired early, and then he left, leaving us to ponder our plans for the day.

Not knowing what the depth of the property was, there was no sense in pulling a tape (we used 300-foot "drag chains" in those days) to find corners. We simply walked along the fence looking for some boundary evidence, we walked until "we got tired" as Ralph told us to. We didn't find any property corners so we took a right, walked out to the railroad tracks, and back to the truck.

We stood there at the truck and discussed our options. It's really not possible to survey a property without some description of the property, like the body of the deed which usually has all of the dimensions of the tract of land. We could have surveyed the "possession" lines, such as the fenceline if we had found it, but there was nothing to survey but the road, the one fence, and the railroad. The party chief said, "Let's go back to Hagerstown and tell them how wacky this old dude is!"

The management in the main office just laughed and said, "Oh well, we can talk to Ralph and straighten the whole thing out." I am still curious about what we were asked to do by the client. This was a good lesson to me; to tell the survey crew what the whole project is about and not just the basic task that is assigned to the crew. Funny thing is, we never did see a deed plot for that property and we never did go back on that job.

A similar situation—only similar in that it was in the surreal world of Ralph Donnelley—was the time we were traversing using stakeout angles and distances that Ralph gave us to stake a property line along a road near his hometown of Hancock, Maryland. That was one of the few times in my surveying career that I was threatened with police arrest!

It started with some unusual directives from Ralph. We followed him to a farm where he announced, "This is our client's home. If you have any problems, come get the farmer that lives here." He then led us to a property corner marked by a railroad spike in the center of the road. Ralph gave us a hand-written sheet listing horizontal angles and distances. He explained that each line of information was for a new set-up and that we should continue along the lines moving the instrument ahead with each new line. Our first backsight was on a fence post, and thus we started a one thousand-foot-plus traverse!

Now if you know anything about surveying, you realize that taking a two-hundred-foot backsight on a post that is six inches wide and then projecting a one-thousand-foot traverse line off of that—well, there's some intrinsic error to that type of survey.

Everything went well until the line departed from the road right of way and crossed a field and into a lawn. We knew something was awry so we went back and started again to check our angles and distances with the same result. The party chief decided to continue and set the point that fell in the lawn alongside a house. We knew we were trespassing, but also following orders.

A gentleman soon came out and asked who we were and what we were doing. He was very patient and explained that his property had a recent survey and showed us his steel markers at the corners of his lawn. He also cautioned us that if his wife came home and saw us we would be in trouble. The man had no problem with us being there but his wife would surely scream bloody murder if she were home. As he was saying this and we were considering our options, we heard him say, "Oh no, here she comes."

I don't know where she had been but she marched around the corner and indeed screamed something that sounded similar to bloody murder! The party chief gave the "pick up" order and I began to gather transit, tape, and hammer and head over the fence. She was already saying the police were being called and that we would be arrested. The party chief apologized and promised her that we were leaving and wouldn't be back.

We drove back to the office and told Mr. Donnelly about our episode and he reprimanded us for not going to the farm to get the property owner. He told us the owner was going to bring his gun to keep that woman away from us so we could stake that property line that ran through her lawn. Once again, the whole job was dropped and we never went back. I wonder if anyone ever paid for that work.

The other thing that made Ralph distinctive was his way of mocking us young surveyor-wanna-bees. When he would lead us out to a job, he would pull his car off the road near the site and we would ask where the property was located that we would be surveying. He would often reply, "Hell, boy, you're the surveyor, you figure it out."

Once he told us to include a particular Mason-Dixon stone in our survey (I believe it was a "crown" stone). I asked where it was located and he gave me his standard reply. I saw the state line sign along the road and realized that the treeline running perpendicular to the road was the famed Mason-Dixon Line between Maryland and Pennsylvania. I asked which side of the road we could expect to find this marker. He just told me I should be able to find it. I began to walk up along the field to the west of the road and he and the rest of the crew began to follow. I assumed that I had made the right choice or he would have stopped me. When I got to the top of a rise, I peered into the treeline and saw the Mason-Dixon stone projecting through the brush. I got somewhat excited and said, "Hey, here it is," Ralph deadpanned, "Hell, boy, you act like you never found anything before."

Ralph did the same thing to me when we were surveying in and around the campus of Shepherdstown College in Shepherdstown, West Virginia. Some previous surveyor had established control points all over the campus, using ¾"-inch pipes with 12-gauge shotgun shells stuck in the top. The dented primer on the shell made a very nice, distinct point to use as a reference. We pulled some distances to find one of these and used a shovel to peel back the dirt. When I saw the well-preserved and stable control point, I exclaimed, "Look at that, there it is," and Ralph flatly said, "Yes, my God, you found it." Everyone had a good laugh at my expense when they heard the voice of an exuberant young man compared with the worn tone of an experienced surveyor who had been a lot of places and had seen a lot of things.

The Fun Party Chief

It was always interesting and sometimes fun working with Mike Sanders. As I stated earlier, we were the main Martinsburg crew that did the fieldwork on Ralph Donnelly's jobs. Mike and I hit it off well because we both liked the same kind of rock music and we both liked to party. There were a lot of times that we worked hard to get the work done to have some goof-off time and have a lot of laughs. When we worked at a frantic pace, Mike liked to call that "busting ass"

Mike would always laugh at my impersonations and antics. I could make him laugh whenever it looked like rain and I would do my "rain dance." I got the nickname of "Chief Pontiac" because I stole a joke that was told by the comedian David Brenner about his big nose. He joked that when he didn't have a costume for a Halloween party, he slicked back his hair and went to a costume party as the front of a '49 Pontiac, and then he would pull his hair back and turn his head to show his profile resembling the Native American, Chief Pontiac, the leader of the Ottawa Tribe.

When I worked with Mike—and some other rodman/chainman that was assigned to his crew—we were always looking for a good excuse to go to the bar. Rain would interrupt the day and sometimes send us home, except, we didn't go home, we went to the bar (more on the bar of choice later). So, as the clouds would roll in, I would do my best interpretation of an old "Merry Melodies" cartoon where the Native Americans (I won't try to act like we were "politically correct," back then we called them Indians) wanted rain. The cartoon depicted a terrible drought by showing large cracks in the earth and vegetables drying up and withering on the vine. Then the medicine man started dancing and chanting, "We-eee wa-aaant ra-aaain." As he danced he stomped and chanted over and over. I would do that same dance with a lot of drama. I also sang the chant with a quavering voice and as I stomped around the instrument. Mike would almost split a rib, laughing so hard! It was so popular that the other guy would sometimes join in and then if it did actually rain, they'd say something like, "Ole Chief did it again."

Mike Sanders with EDM

I liked the nickname, "Chief" even better; it made me sound like a leader, even though I was only eighteen-years-old. The other nickname that Mike tagged me with was, "Aero." That was a derivation of an old nickname of Airwick; which I earned from a young child of a friend, who asked her parents my name. The kid, who was about two years old, tried to repeat Eric and it came out as "Aa-Wick." That was close enough to the famous air-freshener product and so that circle of friends always referred to me as Airwick after that.

When I started working on the crew with Mike, the instrument man on the crew was the previously-mentioned Kent Faughander. Kent had graduated from Frostburg State College (now University) with a BS in Cartography. He had a clear understanding of creating, using, and reading maps; after all, surveyors create maps from their measurements, so this was a good job for him, but he seemed a bit above that work. I think what he enjoyed the most was being on the crew with Mike.

Kent had just graduated from Frostburg and was used to the fun-loving, party lifestyle. I guess Kent felt like he was with the cool crew. He used to wear a ball cap with the letter, "K.I.F" on the front. I asked what the letters stood for; thinking they were initials of a ball team. He evaded the question and I couldn't understand if he was embarrassed or just picking on the "kid" the way my brothers had done. I ventured a few guesses by making up unusual names like, "Kids Into Fundamentals" or "Kings Into Farts." Some of these made Mike laugh and Kent mad. Eventually, I said, "Oh, it's your initials, what's your middle name." Well, then Kent clammed up and tried to ignore us. We kept laughing and asking his middle name and also pointing out the fact that it was kind of weird for a guy to buy a hat with his initials on it. This made him a bit more embarrassed a self-conscience, then the real punch line came to me and I said, "Ooh, I've got it; your middle name must be 'Ingrid'!" This made Mike howl with laughter and made Kent just a bit redder in the face. He had no other option than to simply bow his head and say, "Yeah, that's it."

He tried to join in on the obvious joke at his expense, but we could still sense the sting in the mocking. It was something that I was good at because that was my main form of attack or defense against my brothers. We would continue with the taunts of, "Ingrid" until he seemed to be enjoying the joke as much as Mike and I and then I started calling him, "Kif" and then "Kiffy," to make it sound like something a mother would call out to her little boy.

He didn't like that and he seemed to come up with an idea for a name to call me. I don't know if this name came into his head at that moment, or if he had thought of it before and saved it for a long time—

like ammo stored in your belt—or maybe he used it all the time and I was just now learning this powerful insult! He called me, "Needle." I had to ask what that was supposed to mean, so he went on and told me the whole name, "Needle-dicked, Bug-F***er." Of course, Mike laughed hard at that and suddenly they were the ones with the big horse-laugh and I was a bit embarrassed and defensive. It was all fun because we could just start laughing; especially when in public with other people around and Kent would just say, "Hey Neeeedle." We knew the whole name but no one else around us did.

Mike grew up outside of Williamsport, Maryland, and still had a lot of friends there. His one friend managed a bar in the town, named "Pat's Tavern." Jim was the friend who was usually behind the bar and his mother, Pat, was the eponymous owner. It was a true, small neighborhood bar. The beer was just the standards of the day, Bud, Miller, Strohs, Old Milwaukee, and maybe a Lowenbrau.

All of the patrons were locals who knew each other. I don't believe I would have felt very welcome if I weren't with Mike, but when I walked into the place with him, at least three or four people besides the bartender, would greet him and then usually ask, "Who's your friend?" As soon as they knew my name, I was one of them. This might seem like a bit of an exaggeration or high praise for simple recognition, but there were evenings where I bought one beer and then people would start buying me beers for the rest of the evening. Looking back, I probably should have bought a couple of rounds, but I was a young man who usually only had $10-$20 of spending money each week, after payments on loans.

The interesting thing about Pat's was the food that they had to offer. You could see the smoked herring (known as blind robins), beef jerky, and crackers behind the bar. I believe there may have even been hot pickled sausages and dill pickles in a jar, as well, but then you would hear a patron ask for a "fillet." Mike asked if I wanted a fillet. I had no idea if this was a fish fillet sandwich or something else. I was sure that it was not filet mignon.

It was a ground beef sandwich with some barbeque sauce which was sometimes called, "Sloppy Joe," or as my mom would say, "Wet Wimpie." No one knows where that name was coined because it seemed to me that Mom was the only one to use it! In school, the cafeteria ladies would term that sandwich, which was served on a hot dog roll, a "barbeque," as if we might be eating pork ribs or pulled pork. It seems that Pat would brown the ground beef, put it in a crockpot with some barbeque sauce and then Jim would scoop it out on a hot dog roll and sell them for about $1. That was some good food when you're about half drunk and drinking that cheap beer. We did like to drink beer, sometimes too much.

On Thursday mornings Mike would come into work in rare form. It seems that his wife had ceramic classes on Wednesday night, and Mike would sit in Pat's Tavern for most of the night. Sometimes Mike would pull into the parking space where the work trucks lined up and, jump out of the truck and start singing the opening line of the Led Zepplin song, "Ramble On." He would have a smile on his face and stomp his feet as he sang, "In the days of my youth I was told what it means to be a man." Initially, I wondered why Mike was in such an animated state so early in the day so I asked what was going on. He said, "Aww, I was filling myself with ignorant oil last night." Then he said, "Yep, I was doing the same stupid stuff that I've done since I was a kid, drinking way too much and not getting any sleep." He was fun for the first hour of the day and then I guess the buzz finally wore out and he went into full hangover mode. That was not too much fun!

One day when Mike was not in a very good mood, perhaps due to a hangover, we were running level loops to get elevations on certain vertical control points for mapping from aerial photos. One of the points was on a farm across the railroad tracks from the proposed industrial park (I-81/I-70 Industrial Park) where we were surveying. The point was a "Photo ID", a spot that could be identified on the aerial photo with no known coordinates—just an elevation. I think the point was at the end of a rock outcropping in the middle of a pasture field. As we approached the area with our level run, we realized we had to go on private property within sight of the farmhouse. Mike decided it was more prudent to go ahead and trespass without permission than to stop what we were doing, walk to the farmhouse or walk back to the truck, drive around to the farmhouse and ask permission. Mike said let's just get out there, get it done and get back.

Without going into a whole tutorial on how this type of survey is performed, for those who are uninitiated, let's just say that it takes a little bit of time to sight the rod, take a reading, hold that spot, move the level ahead and repeat. As we got nearer to our target, one of the cows in the herd that had been watching our every move got alarmed for some reason (maybe just the sight of a man with a tripod moving around in what may have seemed like a crippled man doing a weird dance with a three-legged crutch) and it began to moo very loudly in a pulsing rhythm that sounded just about like a car alarm. I joked, "Uh-oh, looks like he has a watch-cow." The other guy said, "No, it's an alarm cow!"

We started laughing while Mike got very nervous. Soon the farmer stepped out from inside the barn, which was about one hundred yards away. He looked up our way and started demanding to know what the heck was going on by shouting furiously. Mike answered in a loud holler across the field, "We work for Fox and Associates. You can call the office if you want - 733-8503" (the office number) and we continued our work as the man yelled some more about getting off of his property. The cow shut up, we increased our pace and got out of there. Of course, the property owner called into the office and Mike got some scolding for not asking permission.

Author and Survey Truck - 1980

Somehow, it seemed that Mike got into trouble more than others. It may have been because of his inexperience, or because he had me and other inexperienced guys who worked on the crew. I think we were all just young and dumb. One day we were assigned to set a few missing property corners on a lot in a residential housing development that had been designed and laid out by Fox and Associates in the past. The boss who did the scheduling, Russ, told Mike to set the corners in the field, if possible. That was a bit more efficient than locating the available evidence and bringing the data back to the office to have the office staff compute the location of the missing corners.

To set the corners in the field, we had to set up on the existing corners, take a backsight on an adjacent corner—which means sighting along the property boundary line—and then turning the angle calculated from the differences between the two bearings of the lines. We were doing just that across the back line of the lot. The trees and brush were very thick, but we decided we could cut the brush, clear the line, and get the job done. Here we were, thinking, "Ha, won't the boss be proud of us." I remembered cutting some small trees, briars, vines, and other non-descript brush, but we got it done.

The next day Mike didn't come to work for some reason. At the end of the day, I was in the office and I got questioned about cutting some ornamental trees. I said, "No way, we cut some brush along the back line but there were no ornamental trees." The property owner had called in and complained about some trees that were cut.

When Mike came in the next day Russ asked him if he cut any ornamental trees. Mike said, "Absolutely not; what kind of trees are they accusing us of cutting?" Russ said, "Silver Maples." "What the hell," Mike said, "Silver maples grow everywhere." When they went to the site, the adjoining owner had cleared away all of the brush and there were eight stumps of Silver Maples all in a nice, straight line along the property line. The owner said he had planted them a few years before, but then left the brush to grow up around them.

The company paid something like $1,500 to have those replaced with fifteen-foot tall Silver Maples and we were given a stern talk about cutting people's trees. We still felt like we did nothing wrong, but we realized that we should have looked at what we were doing instead of going full warrior-style on those trees!

Work was pretty slow as the economy struggled through the winter of 1980. There would be weeks of layoff and I'd have to file for unemployment, then I'd get called back to work for a week or two. This was not good for me; to have too much time on my hands. I would go hunting when there was an open season; in fact, that's when I killed my first deer with a flintlock rifle which I borrowed from my brother Ralph, but I did a lot of drinking and partying during the days when I didn't have to work.

The "Party" Chief

In the summer of 1980 business started to pick up. We were busy with some new residential subdivisions and other survey work. New crews were hired and old trucks were brought out of retirement. We saw some new, young party chiefs with some education and experience start working at Fox & Associates. Tim Witter and Tom Shelly had graduated from Mont Alto and then worked for Draper Sutcliff in Frederick. These two later became partners in a survey business, and then Tom Shelly was tragically killed by a train while walking along the tracks, doing a survey.

One of the young party chiefs who had worked for J.B. Furgeson came to work at Fox and they gave him the old green GMC van that was affectionately called "The Pickle." We'll call him Johnny Stick to protect his innocence. I got assigned to work with Johnny and Tom Shelly one day as the rodman. As we started down the road, Johnny, who was driving, talked over his shoulder to me; asking about my experience, hobbies, etc. Then he asked if I liked to party. I said, "Sure." He pulled out a joint and asked if we wanted to smoke it. Before Tom could even weigh in on it, I said, "Sure!"

We smoked most of it on the way to the job. I guess we did okay on the survey, but when we got back to the office we were asked if we forgot any of the equipment on the site. We looked at each other and shrugged. Well, someone who lived along the road where we were working had called in and said that we left a big knife laying along the road and a kid was walking around playing with it! Sure enough, we had carelessly left a machete laying for anyone to pick up and steal, or in this case, just play with it.

Another time when I was working with Johnny, he was cutting line (clearing brush) ahead of me and suddenly began to run, jump, curse and wave his arms like he was having a spastic fit! As he ran toward me, he ran headfirst into a tree, fell down, and then got up and ran by me, saying, "Hummle farble quank. . . bees." I didn't know what "Hummle farble quank" meant, but I knew what "bees" were. He had just swung into a hornets nest with a machete which caused the hornets to come out en-masse to see who was destroying their lair. He had multiple stings on his head, chest, and arms. He also had a "goose egg" rising up on his head from the collision with the tree. He sat on a stump for a bit and then said, "Take me back to the office, I can't see straight and I can't work." I drove us back and he got in his car to go home. He said, "Tell Russ, I bought it in the head." I thought that was a funny way to express his issue, but I told Russ that and he gave me his typical blank stare, so I explained what really happened. Russ was still perplexed as to why he had to quit for the day.

The Story Teller

William E. "Bill" Fox was one of my favorite bosses. I should probably devote a whole chapter to him, but I never really worked closely with him. He was one of the premier storytellers of all time, which of course means that he was not always one hundred percent truthful. But he always made it sound truthful—well, almost always. Bill had a way of setting you up with his stories. They usually started like typical, normal true stories, then once he had you sucked into it, he would start to exaggerate, just a little at first, then when he really had your attention and some more people were listening and laughing; all of a sudden, he would hit you with a zinger that was pure fabrication.

Here's a good example. Bill talking: (I wish you could hear his voice to get the full effect—it was like a mixture of Leslie Neilsen of *The Naked Gun* and the Penguin from *Batman*), "We went down to West Virginia to do a mountain survey. We pulled the truck over where the property started and looked out the window and there was this wall—I mean to see the property, we were looking straight up."—At this point Bill would launch into a whole side tangent about the mountains in West Virginia and physical demonstrations about the steepness of those hills—as I said, he would suck you into this world of his.—"We decided we had to get up that hill to do our work, so we went down to the hardware store to get some climbing equipment. They didn't have any ropes to amount to anything so we bought six of those plumbers' helpers and some string twine."—By this time, you start to think, *No he's not going to tell us ...* —"We went back up there and tied those things on our feet and climbed right up the side of that cliff!" Yeah, right.

Some of Bill's stories weren't so much lies as they were just strange. One guy that used to work out of town with Bill, when the power company employed him, used to drink too much (typical surveyor). What made this guy unique was that when he got drunk he liked to get everyone awake by yelling, "HELP!" The guys would be out getting their bellies full of beer and then when they came back to the motel this one character would start yelling, "HELP!" Everyone in the motel would wake up and start looking out of their doorways. The rest of the crew would start punching this guy and telling him to shut up or they would get kicked out of the place. The goofy culprit would laugh like a maniac and keep yelling assumedly until everyone was awake. Finally, after being wrestled and pummeled into submission, he would settle down and then start up again as soon as they let him up. After several rounds of this, the guys were hesitant to let him up. They would have pillows over his head and they would be laying on him to quiet him down and he would be laughing, saying softly, "Tell 'em. tell 'em" The others said, "What? tell 'em what?" He would say, "Tell 'em 'HELP!'"

Bill and Bernie had worked together on a survey crew for Potomac Edison, an electric company. Bill told us of a story where Bernie was driving the work truck (a Suburban, I would guess) across a field along the planned power line, and Bill was having a heated argument with the rodman in the back seat. Finally, when the argument was boiling over, Bill dove across the seat and began punching the guy in the back seat. Bernie was yelling for them to break it up because he didn't want to see either of them hurt. When they wouldn't stop, Bernie leaped from the driver's seat while the truck continued to roll down the sloped field, and he intervened in the fight to break them apart. The truck crashed into a tree while all three crew members continued to fight in the back seat.

Bill made a good boss because he had already seen all of the stupid things guys will do in the field. For this reason, he was always preaching (just yelling) at us about what we should be doing and what we really should never do. Before I started working there, Bill fired a guy on the spot for exposing himself to a crossing guard. The story goes that a woman was working as a school crossing guard that the one crew would pass each day. Much like the story about Bob waving at the red-bug lady, they would look for her and she was looking for them and they'd exchange a wave. One morning when the party chief was a bit hung-over, he told the instrument man to drive to the job site. As it was told, the rodman on the crew that day was new to that particular crew. When the instrument man announced that they were coming upon the crossing guard, the rodman put the window down, dropped his pants, and stuck his bare butt out the window, to "moon" the woman. She began to wave and then her jaw dropped. She looked at the truck with much a stern stare as it went by.

When the crew returned to the office at the end of the day, Bill Fox was there waiting to slash some throats! As it turned out, the woman had looked at the truck every time that they went by and knew the name as well as the phone number that was painted on the side of the truck. It also turned out that her husband was a Maryland State Trooper, and when he found out about it, he called the office and demanded that this situation be handled with great prejudice.

Bill asked, "Who mooned the crossing guard?" The instrument man indicated the rodman and Bill asked if this were true. The rodman confirmed the fact and Bill told him that he was fired and he'd get his last paycheck in the mail. Then he asked why the party chief didn't fire him on the spot. The party chief said that he was asleep and didn't know that this happened. Bill told him that if this kind of thing ever happened again, not only would the perverted perpetrator be fired but the party chief who didn't fire him would get fired! I remembered this story and had to fire a guy on the spot when I came back to Fox and Associates as a party chief. –See Chapter "Fox & Associates, Inc. (second time) July, 1987 - January, 1991"

Bill Fox was a smart and shrewd businessman. When he realized that he was spending a lot of money on surveying equipment, instrument adjustments, printing services, and drafting supplies, he opened his store, "Fox Sales and Service." He hired a lady to be at the front desk and sell the surveying supplies and make copies while Bill did the instrument adjustments. I believe he did fairly well with it and eventually sold the business. While he had the store, it was easy for us to pick up any supplies that we needed and just charge them to the surveying firm of Fox & Associates.

One time when an instrument was accidentally dropped—which seems to happen from time to time, no matter how careful one is—we took it directly into Bill's office instead of dropping it off at the store. Bill instructed me to set it down on his desk and as he started to level the head of it, I leaned over to see what he was doing and Bill said, "Don't lean on the desk." I said, "I'm not." He shot a look at me like daggers coming out of his eyes and then glared down at my hands leaning on the desk; something that I had done subconsciously. I felt like a total dolt as I apologized.

He determined that the instrument (a theodolite) needed to go into the shop for adjustments. Bill looked at me and said, "You know you're going to pay for this adjustment, right?" I asked how much and he said, "Don't worry, we can take it out of your paycheck at $50 per pay and we'll see how long that will take to pay it off." He was having fun watching my eyes get wide and probably some perspiration starting to bead up on my forehead; he knew I was getting scared, so he continued, "Yes, you'll see your normal deductions, Federal income tax, Local income tax, FICA and then a special one named "Instrument repair." The other guys were laughing and then I slowly realized that Bill had me. He broke into a big grin and told us that we'd have it back in a day or so.

You must realize something before I go any further here. I have been accused of being a real storyteller and stretching the truth the same way that I am accusing Bill Fox of doing. I will assure you that every story contained in these pages is exactly as I remember it. I have been told many times that I have a great memory and that is the only reason that I can write these stories. Everything that I have written so far happened over thirty years ago!

The Other Eric

There was another Eric that worked with me at that time. We soon became friends outside of work. We were about the same age and he had started at Fox and Associates about four months before me. He worked with Benny most of the time and had become an instrument man before I had. He lived in Chewsville, which is a little village between Hagerstown and Smithsburg. When Eric lost his license because of a DUI, I would pick him up for work and take him home. Since one of my favorite hangouts to see rock bands was the old Mountain View Bar which was in Smithsburg (now a branch bank sits at the location: the southwest corner of the intersection of Maryland Route 64 and Maryland Route 77).

I would often go by Eric's house to pick him up. We liked to get high and drink but Eric liked to get higher, as if high wasn't good enough. We also liked to work together, but Eric was much more introverted when he wasn't drinking. He had a nervous laugh that was more of a snicker through closed teeth but he somehow also managed to wiggle his chin around in a circle while he sucked air back in after each little chuckle. When he got drunk he would talk loudly and laugh heartily. I guess I liked the drunk Eric better, but we are still friends even though he hasn't drunk for many years.

There were other amusing facts about Eric; his dad was a pastor and they lived across the street from the church. One of Eric's friends lived next to the church, his name was Tim Bible, and he liked to get high and drunk more than Eric did. He eventually got a job at Fox and Associates, and I started giving him a ride to work since he didn't have a car, but he was not a close friend. Tim lived a hard life for a young man who had just turned twenty-one years old. His mother told him that since he had a job and worked outside, he should get some new jeans for work. He told her that even though the ones he had were worn out, they would suffice because he wasn't going to live to be twenty-five, due to his drinking and drug use. We were in shock when he told us this, but he laughed like a maniac about it! I couldn't imagine the horror of a mother hearing that from her son.

One night after we quit working together, and we each had different jobs, Eric and I went to the Mountain View to see a band. Eric was very drunk (as usual) and I lost track of him. It was close to the end of the night so I went outside to look for him, imagining that he was so drunk that he wandered off. As I came out the front door, I saw a guy flopped over the hood of the first car outside of the door. I laughed and thought, that guy is passed out from drinking so much and his friends are going to have a tough time with him. I walked around the parking lot looking for Eric and then decided to walk back into the bar.

As I passed the unconscious body which was slumped over the front of the car; now from a different angle, I thought to myself, *That kinda looks like Eric.* Then I did a double-take and pulled his head up off the hood and saw that it was him! I pulled him up into a standing position and asked if he was ready to go home. He was completely limp. I grabbed him around the shoulders and tried to get him to walk; using my beer muscles that I had gained by drinking all evening, I managed to drag him limp-legged for about twenty feet and then I let him slump to the ground.

Nowadays, I would have panicked and called an ambulance. He did manage to slur a few words out so I knew he was semiconscious, but he couldn't or wouldn't walk. I threatened to leave him lying in the parking lot but he didn't comprehend a word of it. Some guys came out of the bar and asked if I needed help. I told them that I would get this guy home if I could get him into my car. They said, "Oh, no problem, we'll get him to your car." I tried to decline but they offered their understanding and explained their experience with other drunks whom they had dragged to the car.

They each grabbed a leg and asked which way we were going. I pointed to my car which was along the side road. They took off like they were taking a dead deer to load into the vehicle and get it home. I followed along warning them that they shouldn't drag him through the muddy ditch, but they just plowed on through, saying, "You want to get him home, don't you?" Well, the answer was yes, but now Eric had his clothes coated in slimy mud and all three of the conscious guys worked like hell to push this limp body into the back of my car.

When loading a dead deer into a vehicle, it's not as bad because usually rigor mortis has started to set in, but Eric was like a wet towel! We got him into the back seat, I thanked the muscle-men and I drove to Chewsville and parked in front of the house.

I told Eric that he was home, time for bed, get out, good night, and everything else that I could say. I had even prepped him by telling him that we were coming up his street and he would have to get out of the car. He could only manage a few pathetic whimpers, like a kid that didn't want to get out of bed for school. I eventually got out of the car, came around to the passenger's side, and dragged him out of the car. Eric slid out onto the grass, covered in mud, and laid there mumbling. I told him that I was driving home and he should either get up or go inside or he would be sleeping in the front yard that night. He mumbled a contented response that sounded something like, "Okay."

I left and drove home without any regrets or guilt. I felt like the old adage, "You made your bed, now sleep in it!" seemed to apply aptly to his situation. The next morning, when I got up and went to my car, I realized how much mud was all through the black interior of my fairly new AMC Eagle. I cleaned it up with much effort and then called Eric to see if he was alright and if he was angry about being left lying in the front yard. He thanked me for getting him home. He didn't care that his clothes were muddy or that he slept in the wet grass.

Eric and I also talked a lot about leaving Fox and Associates because of the layoffs that we endured and the low pay. After work, we would go to the other engineering and surveying firms in Hagerstown and put in applications. One day after I got home from work, I got a call from Davis, Renn, and Schrader, an engineering firm in Hagerstown that used to be called J.B. Furgeson before it was bought out. The man on the phone was the chief of surveys and he was offering me a job for $5.50 per hour to work out of town on Texas Eastern gas pipeline surveys. I was only making $4.75 an hour at the time and he also mentioned per diem was about $30!

This was a great offer and I was ready to accept; the only caveat was that I had to be ready to leave the next morning for Ohio or New Jersey (I don't remember, but it was four hours away). I had to show up in Hagerstown at 6:00 a.m. and be ready to stay out of town for two weeks! I asked if I could have a day to consider it and give my current employer at least one-week notice. He said, "No, I have other guys to call, we need two crews by tomorrow." I declined and hung up the phone. Within five minutes Eric called and told me that I didn't need to pick him up for work the next day. I quickly surmised that we had filled out applications on the same day and that his work application was on the pile next to mine and that he must have received the same offer I had. I asked him if he was going to be working out of town and he said, "Yeah, I took the job." He went on to later work as a field crew member of the Engineering Department of Washington County, Maryland, survey crew after he left the firm of Davis, Renn, and Schrader. We lost touch for a while, but have rekindled our friendship recently.

Payroll Problems

There were several reasons that I decided to leave Fox and Associates. One of which was the intermittent layoffs due to lack of work. I also got tired of the lack of meaningful salary increases. After reminding them of the fine work and decent record of two and a half years, they finally gave me an increase of $0.10 per hour. That put me into a higher tax bracket so more wages were held for taxes. That left me with a decrease in take-home pay!

The other reason that I had for leaving was their lack of cash on hand. For a few months, they were forced to deposit money into the payroll account on each payday. I assume this was a cash advance loan from the bank. We were instructed to go to their bank to cash our checks because they may bounce back from other banks. We all did this as a standard practice for a time, but we all thought that it was very much "bush league" and showed the signs of a "rinky-dink" company.

For some reason, on one particular payday, I didn't stop at the Maryland National Bank, where the company had their account, and I got my check cashed at the First National Bank of Fairfield (long-since defunct), where I had my checking account. They cashed it alright, all one hundred and some dollars and I went on my way. A week or so later, I got a notice from my bank that the check that I had cashed from Fox and Associates had been returned for lack of funds. I would be required pay that money back and they would return my check.

When I confronted the administrative manager, Bill Weikert, with this little financial dilemma, I was told that I should take my next paycheck and cash it at their bank in Hagerstown, take that cash and go to my bank (the nearest branch was in Rouzerville; about thirteen miles and twenty minutes from Hagerstown) to "purchase" the void check. Then I could take that check on a twenty-minute drive to the bank in Hagerstown to cash it and then I would have my pay. I asked who was going to pay for the gas and time spent and was told that was my only choice to get my pay.

I did it and then filled my tank with gas, obtained a receipt for the $20 worth of gas, and took it to Mr. Weikert for reimbursement. When I did so, we had been drinking beer all afternoon on a rain day. When I got back to the office to pick up my car, I decided that I had enough courage to demand some financial restitution. I walked into Bill Weikert's office and told him that I wanted to be paid. After some discussion, I pounded on his desk and said, "I want my money, and I want it now!"

I had learned this tactic from an acquaintance who won a pool match in which he had wagered $20. When his defeated opponent hesitated and made excuses about the bet not being valid, my acquaintance pounded the pool table and made the same demand as stated above. It worked for him and it worked for me. Bill told the secretary to get me some money out of petty cash.

The Beginning of the End

The whole paycheck fiasco was one more straw and it finally broke the proverbial camel's back. When I got a call from Ted at Associated Engineering Sciences, Inc., I interviewed and accepted the job. I was so bitter at Fox and Associates that I told Ted I could start work the next day. He admonished me and told me that if I didn't give them a two-week notice, he wouldn't hire me. I obliged and when I went to Russ to give my notice, on a Friday, he told me that I was done at the end of the day—I was basically fired!

I pointed out that I was trying to do the right thing and that I should have just quit like Eric had done, without notice. Russ told me other people were laid off and he would rather bring them back to work. I got pissed off as I worked that day and finally walked off of the job at noon, leaving the party chief and instrument man to do a topographic survey in the woods, up on the mountain above Hunter's Creek Lake, a part of the Catoctin State Forest.

I walked a mile or so down the road and waited at the breast of the dam until they came by and took me back to Hagerstown to get my car. If they hadn't picked me up, I had already decided that I'd have to walk three to four miles to Thurmont to make a call for someone to give me a ride. I felt guilty about leaving my crew in a lurch, but I also went with them as we collectively complained to Russ about how they treat people. I was done with this chapter of my life and moved on.

I was excited to go to work at another firm and prove what I had learned and how I could be a permanent instrument man. The extra money was good, but I felt that it was a good thing to move to a new set of people and a new survey crew.

Part II

Associated Engineering Sciences, Inc. (A.E.S.I.) July 1981 – September 1983

During my time at Fox and Associates, I was laid off almost as much as I worked. The economy was not good between 1979 and 1981. In July of 1981, I got an offer from Associated Engineering Sciences Inc. to work in the field but also to get more office experience than I had previously received. I got hired as an instrument man and was paid $5 per hour. I was rolling in the dough now! The chief of surveys there was a young, hyperactive go-getter who was two years out of college. Ted had completed the Penn State Associates Degree program at Mont Alto and then went to the main campus to receive his bachelor's degree in civil engineering. He was, at that time, working toward his engineer-in-training license and surveyors license at the same time. Ted directed the surveying work under the license of another man who worked in the office and presumably checked all of the work being done.

Ted and I got along okay except that I didn't have a lot of respect for him because he was all "book learnin' and no experience." He didn't have a lot of respect for me because I was the opposite. He was a driven man and did very well for himself in his career.

When I was working as the instrument man on the crew with Randy Snyder, the firm was awarded a contract from Carroll County, Maryland Highway Department for a realignment of a county road. Ted was going to be the designer on the project, so he decided he wanted to go out in the field on the survey crew to establish the survey baselines. He would be the rear chainman, which meant he would hold the end of the tape on the nails that we used for the "stations."

Normally a nail is set at each 50' interval and then the rear chainman moves ahead after the full 200' tape is pulled out on the fourth nail to be set. When the rear chainman goes ahead to hold the end of the tape on the last nail set, he needs to keep his body off to the side so that the instrument operator can see past him to give line, meaning to give direction for the head chainman to get on the baseline (as determined by the crosshairs in the instrument scope) and set the nail "on-line."

Ted would consistently squat directly over the nail, blocking my view of Randy, who was waiting for me to give direction to get on line. Every time that we started the process of setting a new nail, I would yell at Ted to get off line. He would give me a funny look and question my reasoning. Each time, I would yell, "Ted" and then make a huge, slow dramatic wave to signal that he needed to move his body off of the line.

I learned later that he asked Randy why I kept doing that and Randy had to explain the basics of giving line and laying out a baseline.

After we got the baseline laid out, we had to perform a type of location called "right-angle location." This is an antiquated way of drawing highway plans, which is understood by all the "old" surveyors, but I won't go into detail for those who never had to locate each salient feature along the road with two measurements; a "plus" and an "offset." It's basically rectangular coordinates in the field. Ted kept getting the two terms mixed up and couldn't understand my frustration in trying to give him the measurement that he wanted. He just said, "Well, you know what I mean!" Well, no, I didn't know what he meant unless he asked for the right thing. These are some of the reasons that I didn't get along well with Ted.

There was another job for the Frederick County Roads Department where a bridge was going to be replaced. I guess Ted was going to be doing some design, so he went out on the job site with us to begin the survey. He informed us that we would be back in the "boonies" and wouldn't be able to go out to the town to get lunch. I usually packed my lunch, which consisted of a sandwich, chips, and cookies, but I never had a drink so we stopped at a store in the morning to get a soda to drink with my lunch.

When we arrived at the site, I realized there was snow on the side of the road and I could just stick my can of Coke there to keep it cold until we returned at lunchtime to get it out and drink it with my sandwich. When Ted saw me sticking the Coke in the snowbank, he just shook his head and said, "No!" I wasn't sure why he would tell me that I couldn't keep my drink cold and then he continued, "We don't do that here." I still couldn't understand his protests so I bluntly asked him how I should keep my soda cold. "Oh," he said, "I thought you were just going to throw away your can by hiding it in the snow." I just couldn't believe that his first reaction was to just say "No" without asking what I was doing or why.

As we grabbed the equipment and walked along the road toward our work area a car approached, we got off to the side of the road. Ted acted like a mother hen and starting directing us with his hand toward the edge of the road while commanding like a hall monitor, "Single file." We just laughed at him for being such a nerd and superior.

Ted usually worked in the office as an engineering designer, but he also scheduled the fieldwork. Sometimes when there was a rush job, he would ask me to work with him after the normal workday was over. He would go into the bathroom and change into field clothes and then come out announcing that he was like Clark Kent going into a phone booth and emerging as Superman; only he would come out and say, "Here he is: "Super Surveyor."" I just groaned at that.

One time we drove to a remote area in Fulton County in his car with just some basic surveying equipment and materials to stake out a proposed septic drain field. It was supposed to be simple; just set four stakes with flagging and we would be done. As we walked into the site and found the fenceline along the property boundary that we would use to measure the prescribed distances, we saw large patches of multiflora rose bushes.

Anyone familiar with this invasive species of briar bushes knows that these are a surveyor's nightmare. They grow up to seven or eight feet high and have fiercely sharp thorns that can rip through denim and into the skin. The awful realization came over us that we didn't have any cutting tools! I mentioned the fact and the real possibility that we wouldn't be able to measure up, over, or around these bushes and there was no way to go through them without first cutting a path with a machete or brush hook (which we didn't have).

Ted assured me we were going to get these stakes set because someone was coming in the morning to do percolation tests and needed to have these areas marked out. Well, the only thing that could be done was to whack away at these briar bushes with a wooden stake, that measured thirty-six inches long, but was one inch by two inches in width. It is very hard to cut these types of plants with a sharp machete and not get briars into your skin. It is impossible to do it with a stake and not get ripped to shreds while doing it. That experience still haunts me!

On another occasion of working "after work" with Ted, we had to locate some property corners at a property just below the famous Black Rock—a rock formation on the western side of South Mountain, near the Appalachian Trail, just north of Interstate 70. I believe we had just turned the clocks back in the fall to Standard time and didn't really think about how quickly it would get dark. Ted always worked at a frantic pace. When he did pause to gather his thoughts, it was usually while he was chewing his fingernails and looking around quickly. Then he would start talking fast while he strode along at a rapid pace and everyone with him tried to keep up.

As dusk began to fall, I was having a hard time seeing the plumb bob string to get a sight with the instrument. We were getting close to being done, but Ted insisted that I set up one more time on a nail in the middle of the road! This road, named, Black Rock Lane, is a dead end, so there weren't many cars driving on it at the time, but it was still a little bit scary to set up the instrument in the middle of the road as complete darkness was overtaking us.

I couldn't see the point that I was setting up on through the optical plummet (a small scope that places the center of the angle measuring circle of the instrument directly over a point), so Ted asked the property owner for a flashlight which he used to illuminate the setup point under the instrument. Then he walked to the backsight point and shined the flashlight on the target which was affixed to the plumb bob string. We located the last point as best as we could by flashlight! That might seem like a simple thing, but this was new, uncharted territory for me and it was not easy, nor was it as accurate as I thought it should have been.

I guess I owe Ted a "thank you" for urging me to go to school and learn everything I could about surveying math. While I worked there, the company paid for books and tuition at Hagerstown Junior College (now Hagerstown Community College) for the course offered in Elementary Plane Surveying.

The Tough Party Chief

It was during my employment at A.E.S.I. that I first became party chief, after the former party chief, Randy Snyder, left the firm. When I started there, Randy was the party chief. He was a Mont Alto graduate that lived in the most rural part of Fulton County. Randy was a hunter and fisherman extraordinaire. The only thing he liked as much as hunting and fishing was chewing Skoal and telling stories about hunting and fishing.

Randy Snyder

Randy chewed more snuff than anyone I had ever known. It was nothing for him to go through a whole "tin" of the stuff—or, I mean, "snuff"—in a day! I liked to work with Randy a whole lot. He was laid back and taught me some of my favorite sayings that I still sometimes quote. When things got tough and hectic and I would begin to get rattled trying to get something done, Randy would spit and say something like, "Well, the first thing you have to do is not get excited. Let's stop here, take a dip of Skoal and think about this." I wouldn't suggest tobacco for everyone to calm them down, but it worked for me. It also got me addicted to the nicotine and took the next 16 years to completely kick the habit.

Randy was great though. I used to get a big kick out of his way of saying things that were offensive to most people but just simple talk for him. When I, or anyone else would begin to complain about difficulty, he would simply suggest, “Maybe you ought to toughen up a little bit” or “I think you baby yourself too much.” The best advantage to working with him was that you got plenty of breaks during the day. All you had to do was ask a bunch of questions about his hunting experiences. After a while, Randy would stop, pull out the Skoal, offer it around, and start his story. Then as long as you feigned interest and asked questions, he would go on and on.

I still recite the lines that Randy taught me about being too soft and babying yourself too much. Most people don’t appreciate it, but it reminds me of Randy and how he taught me. It also makes me chuckle a bit to myself; even if the other party doesn’t get the joke.

The Gambler and the Joker

The other member of the crew at that time was Scott Herbert. Scott was the rodman/chainman when I started at A.E.S.I. After I took over the duties of party chief, Scott became my instrument man. We mostly worked together as a two-man crew and laughed like crazy all day. We would always find ways to crack each other up, especially right after lunch.

One of the silly things that don’t even seem funny now—many years later—is the game we would play with license plates. At that time the State of Maryland began issuing license plates that always had three letters at the beginning, so we would try to make them into a three-letter acronym with any kind of silly words being strung together—sort of like I did with the KIF on Kent’s ballcap.

One of our favorite lunch places in Westminster, Maryland, where we were doing a large highway job for Carroll County, was a place that had video games. Video games were very popular then and we would spend more than an hour playing *Galaga, Qix, Ms. PacMan,* and others. After we were done playing and as we were leaving the shopping center to go back to work, we would start looking at the license plates and saying things like, "Jolly Green Giraffes" when we saw a license with "JGG" at the beginning. It sure was silly but we laughed like hyenas then.

One of Scott's other comedy routines was to make up words to songs on the radio or to say old gambler's sayings he learned from hanging out at the horse races with his dad. When the number five was mentioned, Scott would say, "Fever in the flunk house." Who knows the origin of this, but it amused me. Say the number eight and Scott would chime in, "Eight, skate and donate freight!"

Scott had a funny way of saying foolish things right after lunch and it would constantly keep me in stitches. Once while we were sitting outside of a general store on the west end of Hagerstown that sold hardware, and had snacks and food as well, Scott pointed out a guy and said, "I bet that guy's mom would pull him around by his chin when he was little." I looked over and saw this guy with an abnormally extended chin and couldn't help but laugh my head off.

I no sooner calmed down from that one until Scott saw a guy whittling a toothpick out of one of those little wooden spoons that came with Hershey's ice cream cups. I said that was something that my Dad always did after a chicken dinner at William's Grove Amusement Park and they would serve that ice cream. Scott said, "Yeah, now he's gonna go around the corner and whittle himself a little bench to sit on!" I don't know why that struck me as so funny, but I could hardly breathe from laughing so hard. I guess it was just so unexpectedly goofy that it caught me completely off guard.

Another crazy thing that used to keep us entertained between instrument setups and on the ride between jobs was a riddle-like word game. This was a pastime that started while we worked with Randy Snyder. One would come up with a song title and the name of the performing artist and then put it together in a word puzzle that gave hints as to the answer. One of my favorites was the clue for Def Leppard: "hearing-impaired wild feline" or if the artist was Tom Petty and the Heartbreakers, the clue was "small amount and the blood pumper snappers." We even used clues that only we would know, like, "What Ted would say as the Suburban drives up Franklin Street, by Scrawny Elizabeth." That, of course, was "The Boys are Back in Town" by Thin Lizzy.

During those years, I felt like I was making some decent money and found I had a bit more spending money since I was living with my parents and had paid off my student loan. Instead of being financially smart and fiscally responsible, I decided to buy a new car! This was the AMC Eagle that I mentioned before.

My parents weren't sure if it was a good idea, but they allowed me to use the AMC Gremlin that they bought for me to use, as a trade-in. Randy and Scott both questioned why I wanted to go into over $10,000 of debt. They questioned me at one point, "What if you meet a nice girl and want to get married?" I told them that if I met a nice girl to marry, she'd better have lots of money for us to live on. After I reunited with Trudy Short; and we began to date, I realized she was the girl I would marry. She had less money than I did! It didn't matter, we were in love and we got married.

The Loud Mouth

Scott's father, John, who I mentioned above as being a horse race fan, also worked at A.E.S.I. as the head of the testing lab. He was a brash loudmouth who smoked constantly and drank a lot of coffee. On my first day on the job at A.E.S.I., I showed up early and went into the testing lab to see if the surveyors were there.

John Herbert said, "No, they won't be here for another 10 minutes. What's in that Thermos?" I replied that I had lemonade. He said, "F*** that; I want some coffee!" I didn't know what to say. The other guys there said I could wait in the lab, but I said I didn't want to become sterile and pointed at the radioactive symbol on the door—put there as a warning about the nuclear density testing machine used there. John made a disparaging remark about my manhood. I said, "How can you talk to me like that, you don't even know me?" He said—not in so many words—because of my skinny backside, I didn't have the right equipment to become a father anyhow! His exact words were, "You ain't got no ass, I know you ain't got no dick!"

He was a very opinionated character that would share his views on baseball and politics at quite a high volume. He had coached American Legion baseball and could always be relied upon to tell everyone who was sitting around the testing lab, what was wrong with the Baltimore

Orioles. I listened to what he said because I was a baseball fan and wanted to learn all that I could about the finer points of the game.

One time I mentioned an inning in which the Oriole pitcher allowed three batters to reach first and then proceeded to pick each one of them off at first base. Mr. Herbert replied with confidence that the only reason that happened was because they had "little Len Sakata catching" and the runners were trying to get a good lead and steal second base. This seemed reasonable because Len Sakata was not a catcher; he had been substituted behind the plate after the three catchers on the team had already been removed for various reasons (basically a coaching gaffe by Manager Joe Altobelli).

I was interested in this story because I hadn't realized all of the details that led to this unusual inning, but then he went on to say, "That's why the Orioles will never win another championship, because of that little Len Sakata—his arms are too short." I asked why that would make any difference in the fate of the whole team. John said Lenny's little, short arms prevented him from turning double plays and they just couldn't win enough games with that liability at second base. I never discussed baseball with him after that.

Walking with a Dead Man

There was a topographic survey that we conducted in Martinsburg, West Virginia, one winter using just Randy and me as a two-man crew. It was cold and I remember several times telling Randy I needed a warm-up break and he would say, "Sure, in just a minute." I had to clarify to him that I would have to warm up now because I could no longer hold the pencil to take notes (remember, this was before data collectors).

This project site was a H.U.D. subsidized project in an apartment complex. These were low-income West Virginians and there was always some action going on. We saw the police tracking a guy through the snow following a break-in, then later we saw some kid (perhaps the one who did the break-in) trying to sell some jewelry. We had kids in this complex trying to break up our equipment every time we left a tripod set up for a backsight. It was like a little city full of crime!

One day I was moving the instrument ahead while Randy went

back to get the back-sight tripod, I saw some men in uniform motioning and yelling for me to come help. I assumed they were policemen seeking yet another criminal and somehow needed assistance. When I walked toward them and went around a corner of the building, I saw the one man in an ambulance. They were both ambulance attendants (this was also before paramedics were popular). They were asking me to help them carry a dead guy down from the third-floor apartment. I said, "... uh, could you get someone else?"

They said we needed to hurry because his wife and daughter were there, and they were trying to give them some hope that he may live if we rushed him to the hospital since they couldn't pronounce him dead and couldn't really do any work on him in the apartment. The big problem was that he weighed well more than three hundred pounds! I felt sorry for them after realizing their dilemma and offered to hold the doors for them.

There was no elevator, and these guys were struggling down three flights of stairs. I was holding one of the doors, which were at each landing when they came around the corner and the arms of the corpse kept sliding off of his large round belly where they were laid. The arms kept catching on the door frame and they asked me to put them back up. I did but it fell back off and I didn't want to touch it again. I raced ahead to the next door as I saw one of the attendants stuff the arm under the limp body on the stretcher. I was rather squeamish and wasn't much help. They eventually made it to the ambulance with the much-distraught wife and daughter following along behind with sobs and wails. I was glad when that job was finished. I didn't care much for that section of Martinsburg.

Yost, the Host with the Most

While employed at A.E.S.I. we were contracted to survey a tract of mountain land for a man named Yost. He was convinced that he owned much more land than his deed specified. When my co-worker, Scott, and I visited the site and met with the owner, he showed us where his boundaries were by throwing his hands through the air and pointing way over the mountain and off to both sides, saying, "That's all mine." We didn't know what to do. We were used to the owner at least showing us some kind of evidence of their ownership. We saw a wire fence strung on old wooden posts near his driveway and asked if that was, perhaps his boundary. He said, "No it's way over there," pointing far beyond it. Scott and I searched all around without finding any property corners or signs of ownership other than the wire fence. When I told him that there was no boundary evidence in the area where he indicated, he told us there was an old stone wall that delineated his limits of ownership. I asked him to show us where this wall was located because I had searched the area and found nothing.

As he led us back through the woods, he related stories about how he owned a lot more than his deed indicated and that all of his neighbors were claiming land that belonged to him. He also claimed that the previous owners had pointed out various line markings, like the stone wall he was going to show us. I was very familiar with these types of structures that were built in southcentral Pennsylvania and western Maryland by the farmers as they picked the rocks off of their fields and laid them in nice, straight lines with a width and height of about 3 feet; they were typically built on the property boundaries, with some exceptions.

We had walked about 300 feet when he stopped and showed us a small pile of rocks that could have been a natural rock formation with a few more added. It wasn't the typical stonepile that I had seen assembled to mark a property corner; it was more like the beginning of a small stone wall. As I studied it and looked up the hill in the direction of the property boundary, I couldn't see any evidence of a stone wall. Mr. Yost began to get angry as he charged up the hill and said, "Here's some of the wall", as he pointed at small rock outcroppings here and there.

I knew how important it was to interview property owners or anyone with knowledge of the property ownership, but I was beginning to question whether this man was delusional, scheming, or both. I questioned him a bit about the veracity of his claims and this just made him angrier. I finally conceded that perhaps this was the property boundary and we'd have to make some preliminary measurements before we began our traverse and location survey.

After he went back to his house, we measured the deed distances along the aforementioned wooden post and wire fence, and started to find some corners of the fence that matched the direction and distances stated in the deed. As we continued our work and came close to houses on the adjoining properties, I went to the door of these houses and told the owners where we were doing. The first property owner that I spoke with showed us some rebars and caps that were set during a recent survey of his land. These all seemed to match with everything we had observed and measured to this point. He also told us some stories about the arguments he'd had with Mr. Yost regarding the boundary location.

When we stopped to talk to Mr. Yost and told him what we had found, he tried to discredit the veracity of the evidence that we found. Every time I stopped at his house during our survey, he would tell us obvious lies about his neighbors, apparently to discredit the parole evidence that they had provided. The scary part was after he completed each story, he would raise his right hand, as if swearing an oath and say, "If I'm lying, may God strike me dead!" I was always looking around, hoping that God would have mercy on my poor, lost soul.

As we met the various adjoining property owners and told them that we were surveying for Yost, they all had some stories to tell us about how he was a contemptible man. It seemed he had cheated and lied to most of them. Every time he saw us doing our survey, he would stop us and ask about the progress. I was young and cocksure about my work, so I told him how we had found some pipes and pins that were set for corners and other corroborating evidence. Each time he would curse and say, "That's not where the corner is! Those people set those corners themselves—they're not in the right location!" When I informed him that some of them were set by surveyors who used an identifying cap on the top (now required by law in Maryland), he accused the property owners of moving the corners. I tried to convince him that the corners were in the right place,

We completed our survey and discovered that most of the fence and corners that were found, fit into the owner's deed as well as the adjoining owners' deeds. The day that we went to set the property corners, he was angry about where the corners were going to be set. He told us we shouldn't have believed his neighbors, he owned everything right up to their front door. When he insisted we go look at "his" property near the one neighbor's house, we parked the truck and walked toward this forlorn dwelling. As we approached the front yard, a stereotypical mountain man, with long hair, a long beard and a ragged hat, appeared on the front porch with his shotgun and just stood there. He was obviously guarding his property, and I told Mr. Yost that we weren't going up there.

Yost was trying to say the wagon road that his deed listed as the boundary was this old logging road that went behind this "shack." The plotting of the deed aligned very closely with the center of the county road in front of the dwelling. I tried to explain that the deed description was very old and it may have been an old wagon road when it was written. This was immediately met with his vehement disapproval. He was being ridiculous, and I couldn't talk sense to him. He got mad and told us to leave and not come back. We did sneak back to set the corners and fulfill our contract. To provide evidence that the corners were set, we took pictures of them with a copy of the day's local newspaper.

He hired another surveyor to check our work and when that surveyor called Mr. Yost to tell him that he agreed with our findings, Mr. Yost exploded on the phone and said, "You surveyors are all in cahoots!" That surveyor told us Yost had slammed the phone down on the receiver cradle so hard that it didn't disconnect but bounced off and was left hanging so that the surveyor on the other end could continue to hear the outpouring of profanity that Yost was sharing with his family!

I often think about this experience when I give a proposal for a boundary survey and the potential client seems to be a bit on the crazy side. Sometimes, it's more advisable to move onto the next client who seems to have the intelligence and means to accept the survey and pay the fee.

The Testing Lab Men

During the winter of 1982 things were slow in the Inspection and Testing department. Some days we would take the third guy with us from the testing lab just to give them some billable time. One of those guys was Jeff Shetrone. He was from Tomstown; a village just north of Waynesboro, Pennsylvania, that was known for the hillbillies that lived there and their slow way of talking. The town "square" was an intersection with a spring-fed watering trough that the locals called "The Pump", only they said it as "The Paawwwmp."

When Jeff worked with us, he always had his eye out for nice old classic cars; then he would announce his judgment on them. His comment would usually be either: "That's a piece of sled" (meaning that the car was junk), or "That baby's mint"—this of course meant that the car looked original and in great shape or he just liked it.

One day the survey manager assigned Jeff to go out in the field with us on a mountain survey for the State of Maryland, Department of General Services. This was a boundary survey on a tract of land along the Appalachian Trail, on South Mountain, east of Boonsboro, and near the stone-jug-shaped Washington Monument. While we were surveying that day, Jeff kept talking about the wildlife that he hoped to see, including deer and turkeys. At one point, while Jeff and I were gathering the instrument and other equipment to move ahead, we heard a noise that almost sounded like a weak bark from a dog, that Jeff insisted was a turkey. He said, "Stand still here and be quiet and I'll show you how to call in a turkey."

I doubted his abilities and even doubted it was a turkey, but I complied and he began to make a sound with his mouth that was a reasonable imitation of a hen turkey clucking and yelping. The sound that we had heard was repeated and began to get closer. The calling went back and forth for a minute and soon we heard the source of this strange call getting closer and making noise in the dry leaves on the ground of the woods. Jeff was getting excited and intimated that he was calling this thing in.

We hid behind a tree and peered around it to see what was coming, as Jeff called again. Imagine our surprise when we saw a gray fox walking through the woods thirty feet away from us with his ears on alert and his eyes searching for the source of the turkey call! I had to laugh at Jeff who thought he was calling a turkey and was so proud of his ability to make the call with his mouth. Of course, that scared the fox and the moment was gone, but I couldn't help but laugh at this attempt to call a turkey. I reasoned that the fox heard what he considered to be an injured hen that he could quickly catch and eat. Either that, or it sounded like a sick fox and he wanted to see what was wrong.

One year we decided that the survey department (all five of us) would have a little office Christmas party on the last working day before the holiday. Jeff found out about it and tried to invite himself. We told him that everyone was either bringing beer or food and that he wasn't invited. He persisted in wanting to attend and asked if he brought some snacks, if he could come to the party. We allowed it and so he brought a box of *Tid-Bits,* but he kept referring to them as Tit-Bits, and offered them around the group, asking, "Do you want your tit bit?"

We all were feeling good and enjoying the snacks and the effects of the alcohol. After an hour or so, Jeff said, "I should call my wife and let her know where I am." He used the office phone in the same room as us and we heard him telling her that he would be home in just a few minutes. Well, he lived thirty minutes away, so who was he trying to fool: himself or his wife? He stayed another hour and then called her again to say that he was leaving at that moment. He then drank one more beer before leaving.

The whole time we tried to remind him that his wife was waiting for him at home. He just laughed it off. After the Christmas holiday, he came into work and told us that when he got home after the pre-holiday party, his wife had thrown the Christmas tree out into the front yard and locked the doors. She yelled through the door that he could just spend Christmas outside!

We had several different inspectors at different times working on the field crew with me and the others. One of those guys was Jay. He was an older guy with some grey in his goatee and mustache. He was known to have a drink or two after work and he was a comical character. When we drove through town and saw a good-looking young lady, he would always alert everyone in the truck by announcing, "Hello darlin'," but he did it in such a way that it sounded like the opening line in a country song by George Jones. He had a deep voice which made him sound like a country and western crooner. Then he surprised us by showing off his singing voice one day when we were working inside of a school.

The floor slab of this school was settling and we had to monitor the cracks in the walls and the elevations of the slab. We went into a music room where a piano was sitting. Jay stood by the keyboard and started pounding out the staccato beginning of the famous Jerry Lee Lewis song and then he sang, "' You shake my nerves and you rattle my brain!'" Both Scott Herbert and I were amazed, startled, amused, and somewhat worried that we'd get in trouble for playing the piano. Jay let out a hearty chuckle and then moved on to the work at hand.

One of the other inspectors that worked with Scott and me while their work was slow and we needed a third person was Rocky Bishop. He was a big boy with a full beard. He didn't look like the type that worked too hard, but he always called the surveyors "Gravy." It was a snide comment meant to say that we had the "gravy" job, which was much easier than his.

The tough part about working with these inspectors was that they hardly ever worked a full eight-hour day unless they got stuck in the laboratory doing testing. If they went out to a construction site to gather samples or perform inspections, they billed a minimum of four hours. That would allow them on some days to go to two different job sites, bill eight hours, and get back to the office right after lunch. If that happened to several of them, they would sit in the office and play a card game called setback.

The rules of the game are not that important to explain here, but there was money involved which was usually set at twenty-five cents for each time a player was "set" (didn't make his bid) and twenty-five cents for the winner in each game. The game was also popular for the guys working the lab to play over lunchtime. If four or five hands were played over lunch, the big winner might get a jackpot of $2, so it was not really gambling, just fun.

When Rocky worked with us, he insisted on bringing a deck of cards. We had already played some setback in the truck when Randy Snyder had been with us, over lunch, or when it was raining. Rocky liked playing so much that when we were going to the job site in Martinsburg, West Virginia—about thirty minutes, with me driving—he would want to play a few hands of cards. Then when we got on the job site, he wanted to continue. I had to be the bad guy and order them out of the truck to work.

At lunch, I would acquiesce and play a few hands. It always got out of control and we would play for an hour or so. I didn't like it much because I usually ended up spending a dollar or two on my losses. Rocky was such a card player. When the cards were dealt, he would gather his up, shuffle them and rub them and say, "C'mon baby, here we go," as if he could change the cards that he was dealt by coercing them with a rub and soothing words. When he got a good hand, he would chuckle and say, "Now that's right there in my pepper patch." Who knows what that meant to him? I became so enamored with this card game that I taught my then-fiancée, Trudy, to play and we would spend hours playing a two-hand game.

The Other Mr. Oliver

While Randy was still the party chief, we had the "pleasure" of working with the boss's son, Joe Oliver. Joe had been working in the inspection department but had gotten into some trouble when he was recorded on video equipment at a job site performing a concrete test completely wrong. When the owner of the company, Julian (Jerry) Oliver was shown the film and asked if that was one of his employees, he had to confess that, yes, it was and in fact, it was his son.

So, Joe was given an ultimatum: he could work on the survey crew or move out and join the Navy. Joe decided he would try to work on the survey crew. On the first day, he jumped out of the station wagon in which we traveled, and decided he was going to set up the instrument. Randy had to tell him to settle down and he would tell him what he was going to do. Randy was busy finishing reading the newspaper of the day and wasn't ready to get out of the car yet. When Joe got antsy and demanded that we get out and start working, Randy snapped on him. He yelled, "Your name might be Oliver, but you don't sign my paycheck. You better settle down and don't tell me what to do." Joe slinked around the car like a whipped pup.

When we did get out to work, Joe decided he would be like the rest of the crew and chew some Skoal. He put a little dip in his lip and soon was spitting it out, exclaiming that it was too hot, he was dizzy and he might puke! We all howled with laughter and he had to run up the hill to the station wagon to get a drink. Randy just berated him for this.

Joe didn't do himself any favors; he continued to feel the ire of

Randy. As he sat on the side of the road writing his name in a fancy font with keel (a lumber crayon for marking on the pavement), Randy came by, read the name, spat tobacco juice on it, rubbed it with his boot sole, and said, "There, that's what I think of you, Joe!" Then he told him not to waste the keel!

Joe really got the wrath of Randy one day while we were getting cross-section elevations. Joe had the rod and I had the reel end of the tape on the baseline. Joe had the other end of the tape, but kept dropping it and then kept waving the rod around while Randy was trying to read it. I had a hard time keeping him on right angles to the baseline, as well as getting him to hold the tape to get offset distances.

At one point Randy hollered, "If you don't hold that rod still, I'm gonna hit you in the head with a rock!" I couldn't help but aggravate the situation by tugging on the tape on the next shot so that it caused the rod to wave around. Joe didn't want to drop the end of the tape again and get yelled at, but he knew the tugging was causing the rod to wave. He looked at me pleading me to stop and I saw Randy out of the corner of my eye, pick up a chuck of shale about two inches square and throw it at Joe. The rock smashed against the face of the rod about one foot above Joe's head. I was shocked and so was Joe. His eyes got really big as he gasped at Randy. All Randy said was, "I told you to hold that rod still!" Well, as one might suspect, Joe quit and enlisted in the Navy soon after that!

Payroll Problems – Again?

Things went fairly well at A.E.S.I. but the economy was really flat, and the company was struggling with its cash flow. There never seemed to be any money. The company was jointly held by three members of the Oliver family, with Julian (Jerry) Oliver, P.E., being the managing partner. It seemed although we were fairly busy with profitable work, we never had any money for necessary purchases. I wondered if the Oliver family partners; the old man who had commercial real estate around Hagerstown and his son (brother of Jerry) who was an attorney and always seemed prosperous, were taking just a bit too much of the profits for themselves. I never mean to disparage someone who invests money and makes money from their shrewd business practices, but when you force your employees to suffer through poor working conditions, I find that irresponsible. Some examples of what type of suffering we endured:

Using brush hooks and sledgehammer handles with cracks and held together with duct tape, bailer wire (actually rigged together with a wire flag that utilities use to mark underground facilities), and glue because we couldn't afford to replace the handles.

Setting stakes made from yellow pine. They usually had knots and were so brittle that you could only use them in mud or sand. Even if you attempted to pound one into a grassy yard, if there was a knot, it would shatter into splinters.

Being asked to put tire chains on the truck when it snowed because they couldn't afford to put snow tires on the old Suburban! The tires were not just "summer tread" they were "some tread"—not much, but some. I protested and said I would go home before I would crawl around under the truck on a cold snowy morning to put on chains and then risk my life, driving in those conditions.

Having to request money from the secretary for gasoline to keep the truck running so we could go to jobs, bill the client and make money. This seems logical and overstated, but you would think we were asking for a personal loan! Audrey, the secretary would say (in her southern Hagerstown drawl), "You already used all of your gaas money for this month." She always said "gaas" with a long "A" instead of "gas" due to her weird accent. Month was pronounced with a log "O" so that it rhymed with "both."

When we got the Maryland State Highway Administration contract, we needed to operate with five-person crews so they purchased a van—it was a cargo van. Someone went to a junkyard (auto salvage parts) and purchased some seats out of an old Suburban and bolted them to the floor. A plywood box held the tools and materials in the back with no protection to keep them in the back in case of a collision or sudden stop. Imagine riding an hour or two to the Washington, D.C. beltway in the back seat of a van with no windows, except for the windshield, front doors and rear doors, with no air conditioning and no radio. When I drove this, I brought my portable transistor radio and strapped it to the dashboard with rubber bands! We could at least hear *DC 101* with Howard Stern and later *The Greaseman* and the rock music they played on that station.

Then we experienced some of the same financial difficulties that I had seen at Fox and Associates: we weren't getting paid on time. We had to wait a day to cash our checks so that the money that was being deposited on payday would be posted to the account before we tried to cash them.

Mr. Salamone

One payday Ted said, "I have an idea, if you go to this one survey job, the client will give us a check when we show up." He called to make sure that the landowner and client, Nick Salamone; was home. He reminded Salamone that he agreed to pay part of the payment when the crew arrived, and he would pay the rest when we completed the survey. Ted told us that we had to get the check from Salamone so that we would have enough money for everyone to get paid. We drove to the Salamone property in the little village of Chestnut Grove, south of Sharpsburg and near the village of Dargan, which we referred to as, "Dangerous Dargan" because of the various feuds over property boundaries in that area. The Salamone property boundary was similar to what we often experienced in Dargan: the deed had a terrible description and there was very little boundary evidence except for some fenceline near the road.

Mr. Salamone was quite a character himself. I believe his parents came over on the boat from Italy. Mr. Salamone was about sixty years old but looked like he still mended fencelines, drove a tractor all day, and could pitch bales of straw as good as a thirty-year-old. He lived by himself on this farm—more like a farmette because half of the fifty acres were wooded and steep. The deed descriptions on that part of the property called for stumps and "rock" as corners. If the deed calls for a "planted stone," or "stonepile," or "mound of stones," I have had the good fortune of finding those in areas of a mountain that haven't been disturbed by mass timbering or other intrusions, but when it is written "rock," I don't know what to look for.

How big is that rock? Is it a natural outcropping or is it just one lone rock sitting amongst the trees? In this case, where one of the corners was located which called for a rock, we found a lone rock, about the size of a large executive desk, sitting on the ground without any other signs of rocks around it. We felt pretty good about the proposition that this was the corner. When we told Mr. Salamone, we found his one corner and it was a rock, he was disappointed and a bit angry that we wouldn't be setting an iron pin for his corner. We had to explain that this was the original corner and we couldn't set a pin in the rock. He said, "That's not a real corner." He reasoned that just because someone threw a rock at a rabbit and it landed there, we couldn't just claim it was the corner!

We laughed at him and told him how big the rock was. He was still unsatisfied because he wanted a pin set for his corner. He also doubted our abilities and was very upset that we didn't have a compass to lay out his lines. We tried to explain how we do it by assuming north. He asked how we even knew where north was without a compass. I told him we just assumed a direction to get close to north and we always had an idea where north was. He said (with his thick Italian accent), "How de hell do you know where north is without a compass?" I said, "We just look at the sun or watch the geese when they migrate." I made up a false fact that when the geese are returning from the south in the spring, the "V" formed by their flight formation points directly north. He was very amused by this and laughed about it every time we saw him during our survey.

As promised, at the end of the first day on this site, Mr. Salamone handed us a check. We took it back to the office and I took it to the one engineer who managed some of the business. He looked at my hand and asked, "Do you have a check for me?" I said, "Yes, do you have a check for me"—referring to my paycheck because there was no way that I was giving him the check he needed until I got the check that I needed. He pulled out his desk drawer and held out an envelope and offered that I could have it as soon as I handed him the check in my hand. We did a little exchange that included some back and forth offerings until we each handed the check over at the same time. He acted offended that I didn't trust him so I reminded him that the feeling was mutual.

It was about this time that I enrolled in a surveying class at Hagerstown Junior College at the urging of Ted. The firm would pay for the tuition if I got a "B" grade or better. My old friend Mike was also attending at the expense of Fox & Associates. Those were good times! I was starting to learn the basics of coordinate geometry and much more. I had a better attitude than I had in high school, having realized now that this information mattered. I was a step or two ahead of some of the students in that class due to my experience and tutelage by Bob Banzhoff, so I excelled at this. The teacher was a local surveyor named Bucky Teach. He encouraged me to pursue my degree at Penn State, Mont Alto, but I was still paying off my AMC Eagle and didn't want to go into debt again for more college loans. I did manage to earn an "A" in the class and gained some extremely useful knowledge while doing so.

Out of Town – MD SHA

In the spring of 1983, A.E.S.I. was awarded a contract to perform surveys for the State of Maryland, State Highway Administration. I was the only survey party chief when the call came to start the first project in Charles County (in the southern part of the mainland of Maryland). I was told to pack my bags and take a five-person crew out of town to La Plata, Maryland.

The crew consisted of my trusty instrument operator, Scott Herbert, one of his college buddies, Earl Gift, and two kids right out of high school: a basketball standout, Marc Pool, and the equipment manager from the same team, Tim Able.

I worked down there for seven straight weeks (home on weekends) leading up to my wedding. My poor fiancée was left at home to make most of the wedding plans alone. While we were there, we made the best of a bad situation. Just about every night we would stop at the liquor store to get a case of National Bohemian, affectionately referred to as "Natty Boh." This did not make my future wife happy, to learn that we were partying every night, but we were just trying to pass some time between working days.

One of the funniest situations that I experienced in La Plata on that expedition was one night in a Howard Johnson motel room. We had our usual case of Natty Boh, plus some other liquor, snacks, chewing tobacco, and cigarettes. We just sat around consuming all of the above-listed vices and watching TV. Earl Gift, the friend of Scott Herbert, was known to expose himself on occasion because he was proud of being well-endowed. Well, he gets up from his seat, pulls out the trashcan from under the stand where the TV sat, and proceeded to urinate into it—directly in our line of view! He then sat down and giggled to himself for most of the next half hour.

Soon it was realized that the beer and snack supply was getting low. Earl volunteered to drive with one of the high school boys (Marc) to get some more beer. I was against it, but Earl said he was fine and that stuff with the trashcan was just for fun. He said he would be careful and wouldn't have any trouble. I told them to go directly to the 7-Eleven store and come right back because the State Police barracks was right across the road and they were constantly patrolling the area. I should have known better!

An hour later we started to wonder about them. We soon forgot about them in our hazy minds. In another half hour, we heard them at the door of the motel room. Earl staggered in, tossed the survey truck keys to me, and slurred, "Keep those things away from me!" He immediately flopped into bed and passed out, leaving Marc to tell the story of how they drove to a bar several miles down the road, played pool, and eventually left with a six-pack. They were pulled over by a state trooper while they were throwing beer bottles out of the truck. Somehow Earl talked his way out of a ticket. The police officer said that he would let them go only if the trooper could follow him back to the motel without observing any infractions of the law. That must have stressed Earl to the maximum extent, and when he returned without any ticket: the relief must have caused his elation to the point where he passed out!

He later came to consciousness and started to vomit while lying in bed. Someone heard the noise and yelled for help. I believe it was Scott that said, "Come on, let's get him to the bathroom before he chokes on vomit." Some of us ran to his aid and pulled him to his feet. We pushed him toward the bathroom and at one point he stopped us all and said, "Oh yeah, that's just what the f*** I'd do—just push me around!"

After he got into the bathroom and puked for a while, he started to beat on the door saying, "Real nice, guys, you locked me in here!"

We tried to let him out, but of course, he had locked the door. That's the way a bathroom door works; you lock people out, but you can't lock people in. We were drunk and didn't have any good logic in our heads. Besides that, earlier in the week we had had some trouble with the locks in that motel so we believed that he was locked in. We tried to talk him through the process of unlocking the door and he was saying, "I am turning the little button on the knob—it won't unlock!"

I called the front desk and told them our guy was really frantic and we needed to get him out; the goofy guy said, "Do whatever you have to, we'll get maintenance to fix it in the morning." I told everybody that I had permission to knock the door down and to "back off." I warned Earl to get away from the door and then I ran and threw my shoulder into the door. Ouch! It was a steel door in a steel frame. Several more people tried; each one saying, "Get back, here I go!" and then: slam, "Ouch,"—sliding down the door.

Then Scott told everyone to get back. I thought he might get crazy and slam furniture against the door, but I laughed at him as I saw him running toward the door with a shoe in his hand. He slammed it against the door with no effect. Eventually, Earl just turned the knob the right way and the door swung open with Earl standing there with a smirk. Needless to say, the next morning we were all hurting, but Earl was the worst. He puked his Egg McMuffin out immediately after eating it.

We worked on that job while staying in the same Howard Johnson motel, drinking that same swill almost every night. As my fiancée was making last-minute arrangements for the wedding, I was out of town with a bunch of drunk kids.

We didn't just drink; we also played some pick-up basketball against some of the local kids. Marc Poole was a standout basketball player in high school—and went on to be a high school coach when his son was a player—and the other guys were athletic enough that we could compete. It was all fun and games, literally, until Marc went for a rebound and landed on someone else's foot! This caused a serious sprain which laid him up for the next day. Fortunately, it was a Thursday and we were leaving to go home.

Some evenings when it was warm, we brought our baseballs and gloves. I played pitch and catch with some of the guys who were real players, and I soon realized how long it had been since I caught a fastball with some movement. It was fun and I enjoyed the comradery.

Shortly after I was married, we landed another out-of-town stakeout job for the MD SHA. This time we were staking the centerline and right of way. There was already one crew that had been on the job for a few weeks staking the centerline, and this crew included two guys who were married, so I couldn't complain about having to go out of town after just getting married.

These guys, Bob and Steve, were buddies who just graduated from Penn State Mont Alto with their associate degrees and they brought their wives along to stay in the motel with them. The guys on the crew made fun of Steve. He and Bob were both devout Christians, so they did not participate in the after-hours shenanigans that some of us did. I guess that's why they brought their wives along.

Steve made it worse by showing off the skills that he had learned as a church camp counselor. When there was downtime on the job site, he went to the woods and found the appropriate piece of a limb to make a "Woo-Woo stick." You can find a video on YouTube to show you how they are made. The finished product has a propeller that spins when you rub the shaft with another stick. By changing the angle of the stick that is used to rub the main shaft, the propeller spins in the opposite direction. To add to the effect Steve would say, "Woo woo" in a childish voice that made the guys laugh but mostly to ridicule him.

The job was along Route 2 near Calvert Cliffs, and we stayed in nearby Lexington Park, Maryland, at an old motor court. This was the kind of place that had a bar and a restaurant in the front where the office was located, and a looped parking lot with a pool in the middle and the rooms arranged around the outside of the loop. It was summer, so we usually drank beer, ate, went swimming, and then drank more beer.

I remember being drunk and trying to go off of the diving board. I was never really good at diving, but we could get some height and do a seat drop or a cannonball. Then Earl Gift was there one evening giving a boost to the spring in the diving board by hanging on it, under the end, and then as the person on the board would jump off, Earl would let go and give an extra bit of energy to the upstroke of the board. The first time I tried that, I went a lot higher than I expected and oddly hit the water (sort of sideways, as I remember) and got disoriented under the water. As I was taking big strokes with my arms to propel me to the surface, I felt one arm waving through the air. I was sideways in the water at the surface! It was time to get out of the water before I drowned.

After we finished the staking on Route 2, we were assigned to do cross-sections at the interchange of I-95 at the north end of the Washington Capital Beltway (I-495) for a new ramp. This alignment was called Ramp "J", apparently because Ramps "A" through "I" were already designed and some of them built and were found to be inadequate. There were about three abandoned ramps around this interchange and the ones that were being used had tight radii that required the traffic to slow to about 25 MPH to make the transition from one interstate to another.

The alignment of Ramp "J" would allow the traffic to proceed through the transition at 60 MPH. This is what they call a "flyover ramp" design. This sounds wonderful and it really has made this a safer and more efficient interchange, but this alignment went through the ugliest, thickest, steep wooded area within twenty miles of the interchange. If that weren't enough, there were also ground nests of yellow jackets on almost every cross-section line that had to be cleared of brush. I had a five-man crew and Bob had the other five-man crew on that job. We knew how long the MD SHA had given us to do the cross-sections and we knew how many cross-sections there were. We split the section of the job into two and I did some division and determined how many cross-sections my crew had to get done each day. The job was terrible, as described above, so I told my guys that once we got our allotment done for the day—about six cross-sections—we could go to nearby College Park, Maryland to pick up some beer and cruise the campus. The goal was to get these done by lunchtime so we could eat some fast food in College Park, get beer and then hang out in a shady spot.

Bob got the drift of what was going on and felt it was his duty to let our boss, Ted, know about it. Ted questioned me and then laid down the law: we were not to leave the site for lunch! We tried that for one day. The guys complained—they were not happy (neither was

I)—so we went back to our old routine. Some days when the bee nests weren't too plentiful and the brush was not too thick, we got all of the needed sections done by 10:00 am, so I had to get creative to keep the guys under control and not foaming at the mouth for beer before noon.

We created a game where we threw a roll of flagging in the air and the next person, in turn, would try to catch it. If you missed it or dropped it the flagging would spool off of the roll and you'd have to gather it up. Whoever had the most flagging accumulated at the end of the game would have to buy the beer for everyone. We usually drank a 12-pack of Lowenbrau, so it was only about $8 and most of us didn't have that kind of cash on hand, but one guy did.

I couldn't remember his name, but he looked like the cartoon kid, Sherman, the kid of the intelligent dog Mr. Peabody. Since nobody could remember the name, "Sherman," we just called him, "Peabody." He was an engineering student who was doing a summer internship. He seemed to have plenty of money so we would set him up to lose! When it was his turn to catch, whoever was throwing it would chuck it up into the tree limbs. Of course, it would start to unroll as it deflected off of the limbs and away from Peabody as it came down. Then it would roll and he would have twenty-five feet of loose flagging on each toss. When he complained, we'd all defend the thrower and say, "No, it was fair." After about the third day in a row of him being the loser, he complained that he was not being treated fairly. I admitted our conspiracy and he told me he didn't mind buying the beer; he just didn't like being taunted for being the loser!

One of the reasons that everyone picked on Bob was his entries into his job diary. The MD SHA required us to keep a diary of the work that was completed each day. This got handed in with the field books at the end of the job. It was supposed to have information like weather, crew members, amount of work completed, and other conditions encountered. Well, I guess Bob thought it should be like a regular diary and he wrote in it like a junior high girl would do!

We got ahold of it one day and started reading it as a young girl might sound. One of the lines for a particular day was, "It was pleasant and sunny today; a real joy to be outside." We kept saying it in different voices and mimicking a girl twisting her hair. It became such a joke that we laughed at it several times a day. There were also entries about how the crew tried to jump over a creek. The line that made us laugh was, "Terry didn't quite make it over the creek. One wet foot among others." The last sentence has haunted me all of these years later. I don't know what it was supposed to mean or if it was just supposed to be funny, but it never made sense. Of course, we just laughed about it until it was not even funny anymore.

One day Bob found some flower pots with marijuana growing in them (potted pot!) alongside one of the abandoned ramps. They were going to call the cops and start a big investigation. I told them to just let it go but they had to dump them over a bridge just so they could feel like they did their part for the war on drugs.

Since we had so much brush to cut on this job, we had to keep sharpening the machetes and brush hooks. One evening I went to my parent's home and borrowed dad's grinder to sharpen all of the cutting edges, then I touched them up with a file so that the blades were very sharp. I warned the guys on the crew that they were handling some sharp tools and to be careful, but then John Pensinger, one of the rodmen, had a small accident.

We were walking through some high grass and brush that was about waist-high when John suddenly yelled, "Ouch, I'm cut!" He told us that he was just walking with the blade hanging in front of his knee and it bounced off of a weed or something as his knee bent and it cut right through his blue jeans into his knee! I knew I had sharpened the tools to a keen edge, but I couldn't imagine that it would cut with just a small amount of energy. I couldn't believe that he was cut that deep, but when I looked back, I saw him clutching his knee and blood was already staining his jeans from the knee down.

We carried him to the van and put our ice packs out of our lunch (beer) coolers on his leg. Actually, the crew did this while I frantically tried to think about where the nearest hospital was located. This was before cell phones (let alone smartphones) which would have given us directions. I started driving toward College Park and for the first and only time in my life, I followed those blue signs with the capital "H" on them to indicate directions to a hospital.

John was lying on the floor of the van while we discussed the use of a tourniquet versus the use of the ice packs to slow the blood flow. I suggested direct pressure and John told us later that "Peabody" was holding his upper thigh to supposedly hold the femoral artery, but it creeped John out. He also rubbed his leg a bit. In any case, we made it to the hospital before John lost too much blood.

The whole crew sat around in the waiting room for about two to three hours while he got stitches and a cast on his leg. It seemed extreme, but the cast was put on so that he would not bend it and pull out the stitches. It was way past quitting time when he was released with crutches. He was on disability for the rest of the summer.

I was really disappointed in my situation when I realized that the rodman/chainman for the state highway crew was making more money than I was at the time. While the party chief of the state highway crew was talking to me on one of the jobs where we worked in tandem, he mentioned that the consultants made a lot more money than the state crew. I told him that I was only making $5.75 per hour and the rodman in the back seat of their van laughed and said, “Ha, I make more than that.”

He just had to cut brush, pound stakes, and do what he was told. I had to be responsible for four other individuals, manage their expenses, and record all of our work in the field book.

I was married to Trudy Short; that little girl that had a crush on me in high school—and who ran on the same cross-country team as me—in May of 1983 and I was ready to make some more money to support a family. After working at A.E.S.I. for the rest of that summer, I decided that I could make more money working for a construction company, providing layout for highways and bridges. I started checking around for another job and decided it was time to try construction layout for a while; after all, that’s the type of work that lured Randy away from A.E.S.I. with more money. I figured I could do the same and try to improve my financial stability.

Part III

Dewey Jordan
September 1983 – January 1985

I responded to an ad in the newspaper about working for a heavy construction company. The man that I spoke to, Richard Kerslake, P.E., asked me to come by his house for an interview. It was one of the most unorthodox interviews I had ever had! We stood against his workbench in his garage while he gave me the third degree. He was a very nice, quiet, and calculated man. He was a Christian and he let me know it. He offered me the job as an instrument man, making $6.50/hour, as compared to the $5.75/hour that I had been making as party chief. So basically, I was offered a job in which I made more money with fewer responsibilities. The benefits were very minimal, including no vacation or paid holidays, but the idea of an extra $30 per week sounded great.

The project on which I would be working was Maryland Route 32 through Columbia, Maryland. The party chief for whom I would be working was none other than the infamous Clyde Hartzell. This was a guy of whom I had heard lots of stories.

On my first day on the job, I met him at Dewey Jordan's home office in Frederick to get a ride to the job site. Clyde was an older man when I worked with him. He was probably about fifty years old. He admitted to me on the first day that he had been through Alcoholics Anonymous and was now a recovering alcoholic. One of the most outstanding characteristics that I noticed about Clyde was the fact that he always wore shirts with two breast pockets. The second part was what was in those pockets. One held a pack of unfiltered Camel cigarettes, the other had a pack of unfiltered Kool cigarettes. He would alternate smoking them. I didn't even know that Kool made an unfiltered cigarette. I guess it was his alternative to drinking. I verified some of the stories that I had heard about him. All of these stories came from the previously mentioned storyteller, Randy Snyder.

He told me that Clyde was the party chief on a big highway project in the hills of West Virginia. Randy was a young junior party chief who was working under Clyde but had to step up "to the helm" more than once when Clyde was unfit for duty. Clyde had a penchant for drinking a brand of whiskey known as "Old Crow," which he affectionately called "The Dirty Bird." According to Randy, he used to drink it all day long.

Once, when they were staying at some cabins near the job site, Randy was busy changing a handle in a sledgehammer outside of the cabins in the evening after work. Clyde came out of the cabin (drunk, as usual) and told Randy he was doing it wrong. Clyde grabbed the new handle that had been inserted into the hammerhead, grabbed it, picked it up, and started to hit the bottom of the handle with the blunt end of an ax head. He swung the ax as hard as he could, one-handed, and eventually missed the handle, and busted the back of his hand, which was holding the hammer. Several of the people present at the time immediately expressed concern about the wound as the blood was running from the lacerated gash on his hand. Clyde quickly pulled out his old hanky (which is more properly called "bandana") and wrapped his hand and continued whaling on the handle. Randy said, "Hold on, Clyde, I got the idea; here, I'll finish this." Clyde questioned why Randy was so concerned and Randy simply pointed at the badly-wounded hand. Clyde looked at him and said matter-of-factly, "Oh, I just knocked the bark off of it!" and then finished the job before going back to finish his bottle.

Randy said that many times when they were going out for supper, they would ask Clyde if he wanted to go along, and he would say, "No, just get me some burgers and a big bottle of "The Dirty Bird." He would relish the new bottle when he broke the seal with a twist of the cap and then tilt it up to his lips and gulp a couple of swallows. I think it's hard for anyone who has tasted whiskey to imagine doing that regularly. The burning in the throat makes my eyes water at the thought.

Once during their work together, Clyde was unfit to take the notes and calculate deflection angles, as they were laying out a curved baseline, so he delegated those duties to Randy while he gave line through the instrument. Clyde was shouting directions while bent over, looking through the scope. Due to his inebriation, he lost his balance and leaned against the tripod, pulling one of the legs out of its secure anchoring in the ground. The cadence of his vocal commands (i.e."Right a couple feet ... right a foot ... right a couple tenths ...") was interrupted as he caught himself, grabbed the leg, and stomped it back into the ground. Then he continued with the same patter: "Now left a couple feet ... left a foot ..." Randy interrupted him abruptly and told him he had bumped the "gun" and needed to re-level and get a new back sight. Clyde told him everything was fine; he had stomped the leg back into the same hole. He thought it insignificant that the instrument was out of level and all orientation was lost!

Even though those Clyde Hartzel stories were second-hand, I felt I had a right to tell them since I worked with both of the parties involved and confirmed their accuracy, at least, with both people. I think Clyde was so hazy on most occasions during that part of his life that he couldn't remember the details that Randy had given.

My experience with Clyde was short. After two weeks, Dick Kerslake decided that I was better suited to be the party chief and Clyde could go back to the project in Hagerstown to be the instrument man. I was assigned the company truck and began taking the responsibilities of laying out slope stakes on the new section of highway where the earth-moving scrapers were getting started stripping topsoil. I worked with an inexperienced rodman named Steve that I tried to train to operate the instrument. It was difficult working with him and learning so much new information about the construction business.

The Misfit Crew

I was soon introduced to my new boss, Jim Miller. At the time when he started, we had two, two-man crews on the project. The other party chief was a man by the name of Meeks. He and a boy by the name of Joe LaFonte worked together on the east end; from US Route. 29 to Interstate 95, while Steve and I did the west end; which started at the tie-in to old Route 32 near Pindell School Road and stopped just short of Route 29. Soon after Jim Miller started, Mr. Meeks was not happy. Jim started giving orders and telling us where to go and stake the roadway regardless of the location. After several weeks of that, Mr. Meeks left and I became the party chief of a three-man crew that was responsible for stakeout on the whole of both projects.

After several months of working with a misfit crew that barely knew the basics of surveying and laying out bridge abutments and piers, it was determined that Steve would do better driving a back-dump Euclid. This is a huge off-road dump truck, used primarily to haul rocks. It is physically abusive as the rocks hit the tail end of the truck when it dumps and the whole truck bounces and rocks. The job is also mentally abusive with boredom. The "straw that broke the camel's back" for Jim was when Steve had set a wrong angle in the theodolite to stake out the bridge wingwall. Jim asked Steve to check the angle and Steve retorted, "I don't have to—I'm right." When I checked it, the angle was ten degrees off, and Jim told Steve he should be driving a truck and not surveying. Jim had no problem with relegating Steve to this type of mindless work; he felt it was better than laying him off. To that, I agreed, and I was glad to be rid of him.

Jim soon found a replacement by hiring a kid that had worked for him in a motorcycle shop that he used to own. This boy named Mike Chinea was about nineteen years old and had already served time in jail for driving violations left unpaid and missed court appearances. Mike and Joe hit it off immediately. They had something in common—they were both dope heads. It was interesting working with those guys and trying to trust them with anything.

These kids would show up some days acting slow and tired, and I would eventually find out they were hungover from drinking alcohol and/or smoking various substances. Other days I saw them full of energy and excited to be at work. Later when I asked them about it, they had no embarrassment in telling me that one of them had some cocaine that they were snorting off and on throughout the day. They even told me about several incidents when they were working under the influence of LSD and PCP I had my own experience with these drugs that I won't go into detail about, so I knew a little about what they were going through. I was married and becoming a father for the first time during this period of my life, so I struggled to stay on the "straight and narrow" and be a leader and mentor to these guys.

At some point, for one reason or another, Jim decided to hire another stoner with a prison record, named Jimmy. He was a friend of Mike's and so the drug drama continued, now with three guys under twenty-two years old and as wild as any guys could be.

On his first day, Jimmy was quickly and abruptly introduced to the world of construction workers in the form of our pipe crew foreman named Bobby. Bobby was a "rough-and-tumble" kind of guy with a foul mouth, a clever mind, and a hyperactive disposition. As soon as he spotted Jimmy, Bobby came over to him and asked a pointed, rude question about Jimmy's previous sexual encounters with African-American females. It had a lot to do with the old fabled proof of manhood relying on "splitting a black oak." The comment was rude and completely unacceptable to me because it contained the "F" word and the "N" word, and harsh in the way it was delivered: close up and eye-to-eye.

Before Jimmy could answer, Bobby continued on with comments about his long hair and Harley-Davidson earrings and how nice he would look in a dress! He told Jimmy that if he put on a dress and wore those earrings with his nice long hair, he'd take him out dancing on Saturday night. He finished by noting his backward baseball cap and told him, "You better get that hat turned around and figure out which way you're going here this morning." Bobby was quite a character. I would be remiss and neglect the whole purpose of this book if I didn't relate some of his stories.

The Pipe Foreman

Bobby would come into the field office (which was an abandoned house that was left standing in the middle of an exit ramp on the proposed highway) and start looking over the reduced plans that were taped to the walls. He would start thinking out loud and saying things like, "Where's that long run of pipe?" Being the helpful guy that I am, I would leave the cut sheets or whatever calculations that I was working on and walk over to the wall with the plans. I'd help Bobby find the run that he was supposed to install and then I'd find the size of the pipe, the length of the run, what type of end treatments were designed (i.e. inlet, endwall, rip-rap, etc.), right down to the size of stone rip-rap. Bobby would be writing all of this down on his notepad.

When I would tell him the number of rip-rap, which is a way to size the stone from the quarry, he'd always look at me quizzically and ask, "That's the big shit, right?" I didn't know, I was just a surveyor! He relied on our help a lot and always called me an engineer. I would try to correct him, but he would say, "Well, a field engineer." Whatever!

Whenever we would stake out a run of drain pipe, he would go out to the site and look at our stakes. Many times, he would come back and say, "Something doesn't look right, you better go check that." He was never specific, but I got the feeling he knew something that I didn't so I'd go check. I never found a problem. After about three or four of these wild-goose chases, I would press him for details. "What looks wrong about it?" I'd ask. He would say, "Just check it for me." A few times if I resisted, he would offer to buy me a coffee if I would just go check it. I eventually realized that he either lacked the skill to check my work or just didn't want to take the time, but somehow felt that my work needed to be checked.

One of my favorite incidents about Bobby was the time when the management hired a new track-hoe operator for the pipe crew. The track-hoe is sometimes, more appropriately, called an excavator. It's one of those big machines with a bucket on a long hydraulic arm that is used to dig deep, wide trenches through any type of material except solid rock. This new operator was digging when we arrived to stake out a pipe run.

Bobby told us the guy was no good, didn't know what he was doing, and wouldn't last a whole day. Bobby had told the guy to dig a hole for the inlet, and the guy was cautiously probing around in the dirt with the bucket. Bobby was cussing and saying, "The guy can't even dig a f***ing hole!" Soon the operator had quit digging and shut the engine down. Bobby went wild and screamed at the guy that those machines were started and shut down only by the mechanics and he'd better start that thing back up and continue digging.

The operator wanted to know how deep. Bobby told him to dig until he was told to stop. We returned to our discussion about the pipe and soon the hoe was spun around, and the horn was sounded. Bobby didn't even flinch, he simply looked at us and said, "That guy is gonna be gone before lunch." I asked if he would fire him and Bobby said, "Oh no, he'll quit—I'll make sure of that."

About the time he finished saying that, the horn sounded again and it was as if that horn were a starting gun for a 100-yard dash that only had one entrant. Bobby sprinted toward the machine, leaped up to the cab, and began screaming loud enough to be heard over the roar of the diesel engine. "Look here, you don't beep at me—I beep at you! I told you to dig, so DIG." We were quite amused by the display, but even more so when we saw that operator walking down the road with his lunch box about 11:00 a.m.! Bobby had accomplished what he set out to do.

Been Everywhere but the Electric Chair

Back to my crew and my boss, Jim Miller. Jim was such a cool character. His main duties were to coordinate the sub-contractors and then also coordinate our stakeout with the grading foremen. I guess Jim was more of a mentor to these guys than I was because he was more permissive and therefore more "cool."

Jim used to tell outrageous stories about his past and every one of his stories would make the boys go, "Oh no, that must have been terrible," to which Jim would undoubtedly insert, "Ah, that shit don't bother me, man," and everyone would emit nervous laughter and feigned respect for Jim.

Away from him, the boys and I would laugh about his unbelievably dramatic stories and how he acted so macho throughout it. For example, one of his stories involved someone breaking into his van or something, and Jim sneaked up behind him, confronted him, and beat the guy to within an inch of his life. The whole time he was telling the story, he would make the meanest facial expressions and tell how angry he was. It was a good show, but after a while, we were sure we were being fed a line.

Another one of Jim's specialties was in getting out of work by loafing. Jim had to travel across about ten miles of the roadway under construction to coordinate the subcontractors. The east end of his domain took him into a small town just north of Laurel, Maryland, called Savage. In this small town was a small restaurant named "Ma's Kettle." It specialized in serving greasy food of every type. This was Jim's second stop of the day after sending the workers out to do some stakeout. Sometimes he would take me along; other times he chose one of his “stoner boys” that had somehow found his favor on that particular day.

Jim would order up the full grease platter that included sausage, eggs, and homefries. These home fries, which are just fried potatoes and onions, were really something! I used to watch them throw these things on the back of the grill and cook all morning. I still have a hard time trying to digest anything like that first thing in the morning. Jim was really in his element sitting at the counter, smoking a cigarette, drinking coffee, flirting with the fairly unattractive waitresses, and pontificating on any and all subject matter.

Many times, when we got back into the truck after our hour-long breakfast break, the two-way FM radio would be crackling with the voice of the field superintendent on the job site—Frank. Frank was from Virginia and had a unique southern drawl that was usually saying, “DJ-4 to DJ-9 (call letters from Frank to Jim) where are you, Jim?” Jim would make some lame excuse about being away from the truck (it was the truth) and he was on the east end of the job (again, true). It was interesting to see how someone could be so insubordinate and still get away with it.

Jim loved to goof off. He would sit in the field office, smoke cigarettes, drink coffee and read magazines whenever he got tired of driving around. He always had something cute and clever to say for any circumstance. If someone tripped, he would immediately say, “Paint that orange for you?” If someone showed him a new or unique item he would quickly state, “Guaranteed not to chip, dent, rust, bust or collect dust.” This was said with such rapidity that most people never understood what was being said, but they were expected to laugh just the same. If Jim was driving behind a slow vehicle he would say (usually both, consecutively), “Safety first—a little left rudder” and, “Gas pedal’s on the right.” Every time that Jim heard a horn beep, he would say, “Go ahead, we got a ride.” If things were particularly slow and he was simply sitting at his desk and smoking, he’d look wistfully into the corner where the ceiling meets the wall and say, “I asked (pronounced, ‘aksed’or ‘axed’) myself; should I, or shouldn’t I—then I couldn’t think of an answer.” I think I was supposed to laugh at that too, but I got tired of it and ignored him.

One thing that Jim taught "the boys" while he was telling one of his stories was how to defend oneself if a person grabs the other person's forearms during an altercation. He made it sound so fascinating and useful that we would sometimes practice it on each other. One time, while I was getting ready to go out of the field office with plans and field books under my arm, Joe decided to grab my arm and initiate the "move." When I told him, I was busy and in no mood to spar with him, he simply grinned, maintained his hold on my arm, and said, "Come on man, aren't you going to do the move?"

I was mad and impatient with him, so I quickly dropped the plans and books and grabbed his hand, twisting his wrist and pushing him toward the floor. This was how the "move" worked; it was the use of leverage to force a person to their knees. As Joe was going down, he gave me a quick jab punch to the jaw. I increased my pressure on his wrist and shoved him the rest of the way to the floor. As I drew back to administer my punch to his face, he quickly and repeatedly apologized and begged for my forgiveness. Being the nice guy, I let him up and put some ice on my lip that was cut inside and starting to swell.

I felt magnanimous about forgiving him and letting him go until I heard him bragging to some of the grade foremen about the fat lip that he gave me and that I "… didn't even do anything about it!" I got mad again and threatened to finish the fight right then and there, but the other guys talked me out of it. There were several fights between me and Joe and Mike. I won't go into details because most of them were scuffles with lots of wrestling and only a few punches thrown in.

At some point, Joe's time with us was ending because in his words he "wasn't into surveying anymore." This was interesting since he later decided that he didn't really mean that, but his work showed otherwise.

Jim told me if I wanted to get rid of Joe, I'd have to do the firing. I was a little bit concerned about firing Joe because I knew he was a hot-head and once while he was drunk, after work, he said he would punch any boss that fired him. Then he looked at me and said, "On my last day when I leave, I'll punch you right in the face." I couldn't believe he would say something like that to me while I was standing around drinking beer with the guys and laughing.

So, on this day I decided I'd better just tell him to leave and get it over with. I remembered the foreman on the bridge crew telling me how this little, short guy who was the foreman of a bunch of very large men on a dam construction site would fire the ones that he needed to get rid of by grabbing a short length of steel reinforcing bar and holding it behind his leg while he called the intended victim over to him. He would then hold it up behind his head in a threatening manner and cursing the man and tell them to "Get the hell out of here! You were never worth damn anyhow!"

After Jim and I had our talk, I went into the field office, poured a very hot cup of coffee, and told Joe we didn't need him anymore. It wasn't like telling him to get out or anything—I just said that we didn't need him. I held the coffee cup ready as he walked past. I was ready to dump it on his face and punch him if necessary. He simply walked by with his trademark toothy grin and then left.

Flirtin' with Disaster

The bridge crew was a different bunch. There were some strange characters. Most of the crew was fine, but this one laborer was strange. It was known by my stoner crew that the bridge crew would smoke pot or sometimes PCP during their lunch break and then go up on the concrete pier or steel beams to work. This was a big bridge that was being built over the Little Patuxent River. It was seventy feet above the water and we had to be lifted to the top using a skidpan on the crane. It was a little nerve-wracking for me, and I guess everyone that wasn't used to the height.

One day this strange laborer was standing next to me on the base of the pier cap that we were laying out. We were waiting for the instrument to be lifted up to the platform where we were standing. He was leaning on the handrail next to me and I was enjoying the warm sunshine of a nice day. He started a strange conversation with me by asking, "You know what everyone has?" I tried to come up with something witty, but he said, "No, it's contact—we all have contact." When I looked at him quizzically, he referred to the platform we were standing on, saying, "See if we didn't have contact with this platform, we'd be falling to the ground there." As he said it, he grabbed my arm and pulled me toward the edge. Whenever you are that high above the ground, something like that makes you feel instantly nervous and a bit scared. I didn't know what this guy was up to, but I didn't trust a dopehead at that height when he was talking crazy.

Then he said, "What's something else that everyone has in common?" I told him I didn't know (and really didn't care to know—but I didn't say this). He told me that everyone had the love of money. I disagreed, saying that I didn't love money; it was just necessary for life. He laughed and said, "I love money, that's why whenever you fall off of this bridge, I'm going to take your wallet and spend all of your money." I got mad, pulled out my wallet to show him the few dollars that I had in my wallet, and said, "Oh, that's real nice, you would pick over my crumpled, bloody, dead body just to get a few dollars, and some pictures of my family." I walked away from him and never got near him again.

Maryland, Route 32

Riding the skid pan to the top of a scaffolding built around the pier cap or abutment concrete structure was one thing, but walking on the steel girders was something else—something I never wanted to do but had to do as part of my job.

When Dick Kerslake told me we would be shooting grades (obtaining elevations) on the top flange of the steel girders while they were set in place on the top of the concrete structure, I gave an emphatic "No." Then when they told me that there would be no harnesses or nets for fall protection, I gave a strenuous negative answer, such as, "No f***ing way." He and Jim Miller both assured me that it would be easy. There were people there who did it all the time and you just had to get over the psychological part—which I deciphered as "You just need to be crazy to do it." I guess it was like my Dad used to say, "It's just mind over matter—if you don't mind, it just doesn't matter!"

Dick offered that when the steel was delivered, we could take a look at it and they would set up a ladder to get on top of the steel and I could walk on it while it rested on the ground. I couldn't imagine how this was going to work, but then he pointed out that the steel girders were eight feet high and were built on a one-degree curve. In case you're interested, that means that they were curved with a radius of 5729.58 feet.

When the steel was delivered and set off on the ground near the bridge under construction, we went there and I climbed up on the top of the steel girder, a safe eight feet off of the ground. The flange (top part of the girder) had varying widths: at the abutment, it was twenty-two inches wide; then it stepped in to eighteen inches wide and finally, in the middle of the span, it was fourteen inches wide, then it stepped back out as it went toward the part that rested on the pier, where the flange was twenty-two inches wide again. Well, I could walk on that steel like it was a sidewalk. I started turning around on it by spinning on my heel and it was easy. I commented that it was easy when I was only eight feet above the ground, but it would be different when it is seventy feet above the river. They told me the only difference was in my head; it would be the same steel, just in a different place.

They also coached me on not looking directly at my feet because when you did, your eyes could be drawn to the flowing water (or in some cases, traffic) under the steel because of the movement just next to your feet. I practiced on the steel when I had a chance and sure enough when it came time to get up on the steel when it had been placed on the concrete structure, I felt confident!

Walking the flange of the steel was not too bad, but connecting the steel girders were four-inch-wide pieces of angle iron with other bracing called diaphragms. If we needed to go from one girder to the other, we could step across on these diaphragms instead of walking back to the abutment.

Stepping across those 4-inch-wide pieces of steel required some nerves of steel. We had to place our foot out on it to feel our balance and then quickly and smoothly take one more step and then onto the other girder. I watched while Joe paused, and then stutter-stepped, one time while doing this and I realized that he was most likely high on some illegal substance, so I told him to get back to the abutment, sit down and stay there. He tried to tell me that he was alright, but I said, “NO, I don’t want to watch anyone die on this bridge,” so he listened and sat out for the remainder of our time on the bridge that day.

Time to Move On

While I was working on this job my wife Trudy and I had our first child, a son. We named him Dustin. I had told Trudy to call the main office if she went into labor and they would call me on the truck radio. I was afraid that I wouldn’t make it to the hospital to be there for her; after all, we had taken the Lamaze classes and I was ready to be a birthing coach. In the end, she got an appointment for induced labor and we soon had a ten-pound, four-ounce baby boy!

Trudy was worried about me every day that I had to work on the bridge and walk the steel beams. She told me she prayed every day for my safety so that she would not be left a widow with a small son. These were some scary times for all of us, but I put on a bravado that seemed to work; I just didn’t worry but remained cautious and calm. The good thing is, I survived and learned a lot about the deflection of steel while under the weight of the concrete, reinforcing steel bars, and the steel deck pan which held the concrete. I often say, “I’m glad that I did it, but I would never do it again!”

When the bridge and most of the main infrastructure were finished on that section of Route 32, we moved to another bridge and a short section of roadway about four miles away on Route 108. That was a small job and when that was finished we were out of survey work, so they asked me to run a trenching machine. I was not interested in the least!

I have always disliked operating machinery. I don't even like a chainsaw or lawnmower. The noise and dirt are part of my reasons, but there's something more to it. I almost feel like the machine is working for me, but I'm also a slave to the machine. Well, anyhow there was other work for me.

Jimmy and I went to the I-795, Northeast Expressway project as laborers to shovel rock dust and lay electrical cables. It was physical work, during the winter and was not what I was looking for in my career. In fact, on several occasions when they sent us home because of snow, Jimmy and I canvassed his neighborhood, around Randallstown, where there were some wealthy folks, to shovel snow just to make a few bucks. Some days we made more shoveling snow than we would have made shoveling rock dust! I very quickly sent out resumes and pursued other options.

Part IV

Rettew
January 1985 – July 1987

After submitting resumes to firms all over South-Central Pennsylvania, I was interviewed by a firm in Lancaster, Pennsylvania, named J. C. Engineering, Inc. I was interviewed by the president of the firm, George Rettew, a.k.a. Hank. The company had a small office in downtown Lancaster with about fifteen employees. Two crews worked in the field and they wanted to hire an experienced person. They made me a good offer of pay and the benefits were very good. This firm grew quickly and soon adopted the name of the owner with controlling interest; hence the name changed to Rettew partway through this engagement.

In my interview, I also met Chief of Surveys, Ron Beam. He had been working in the field that day and came in wearing his LL Bean rubber-bottom boots and a huge grin within his beard. He was a thin, wiry guy who chuckled and asked some tough questions while squinting his whole face into a grin. I started working at the firm shortly after my 24th birthday.

On the first day, I worked with Doug Wolfe (Wolfie) and Ron Horst. These guys were very nice, Christian men. I figured they were both a little older than me. I found out that Ron was twenty-seven and Wolfie was thirty. He still looks five years younger than his license shows. These guys were very helpful in getting me acclimated to the culture at J.C. Engineering. I was soon happily surprised and relieved to know that this firm treated its people with the respect that a professional deserves. I hadn't experienced that with Dewey Jordan—they tended to treat all of the employees as workers and nothing more.

I soon started working mostly with the two licensed surveyors: Ed Warfel, who did most of the scheduling, and Ron Beam. They were both very good at what they did and they taught me a lot. Ron Beam was an exceptional mentor and took a lot of time teaching me about boundary resolutions. I was anxious to be the party chief and show them what I could do, but Hank kept telling me to wait and I would prove myself in due time.

I was happy with the pay raises that I received, but I was not happy working under another's direction every day in the field. It's difficult when one is used to taking a break and eating lunch when he is ready, and then one is forced to work within another's personal preferences. The worst part was waiting to eat lunch—with Ron Beam being the worst.

I would get hungry about 10:30 a.m., and I could hardly wait until lunch. Some days we worked on a project site near a potato chip plant. I could smell that hot grease and potatoes frying and my stomach would growl. Sometimes when we were doing stakeout, I would begin to look forward to our lunch break by 11:00 a. m. About 11:45 a.m., we would finish a set-up and Ron would tell me to pick up the instrument. I would think, *Oh boy—lunch,* and start to mentally pre-taste the sandwich that I had packed. Then Ron would say, "Go ahead and set up over there." I'd look at the stakeout sheet and think, *Okay, this is a short set-up so we can finish this shortly after noon*. We'd get done that set-up by 12:15 and, again, I'd hear the order to "pick up," then to my great dismay, he'd tell me to set up the instrument on another control point! About 12:45, Ron would say, "Let's take a lunch break." To me that meant shoving food in my face so fast I'd be sick—but to him it meant sitting in the truck, listening to the radio while he smoked two cigarettes and drank iced tea (in the summer) or a cup of coffee.

The Funny Guy from New Jersey

It was only a few months until I was given an assistant to work with. His name was Chris Embry. He had a little bit of experience surveying in his home state of New Jersey. Things were a little bit different for him in Pennsylvania—he was not used to hills or rocks.

Chris met his wife while going to school in York, and had married and moved to Lancaster. He was one of the funniest guys to work with! He had lots of stories from his past life in New Jersey, where he had worked at one of the casinos in Atlantic City and had worked for an awning company that installed awnings all over the area.

Chris related all kinds of stories about gamblers, welders, and a chance meeting of Susan Anton. His notion of the once-famous actress and wife of Dudley Moore, was, “She has a big butt!”

One of our favorite things to laugh about together was the old TV show, *Green Acres*. Chris would watch the re-runs on cable TV and then come in and remind me of the shows I had watched during their original runs in the ’60s. We would even do some of the dialogue from Mr. Haney, Mr. Douglas, Hank County, your Kimball agent—or, I mean, Hank Kimball, your County Agent—no, that’s not it. Oh well ... People would think we were weird and that made it even funnier.

There was an episode in which Chuck, Mr. Douglas’ nephew, came to visit. He was a hippie (or beatnik) with long hair and a goatee. When Chuck went out to stay with Eb in the hayloft of the barn, Eb tried to be a good host and offer to hang up his coat. When Chuck said, “No, no thanks.” Eb looked at his head and since there was no hat, but plenty of hair, he asked, “How ’bout your hair?” This became our amusing and popular saying to greet someone with, considering that one of the guys in the survey department was bald and another had shoulder-length red hair.

This was one of those inside jokes that I continued to repeat, even when the present company did not get the joke. It was almost like a sickness and caused people to wonder what I was talking about when I repeated these lines that caused me to laugh. Who cared if they didn’t get it?

Chris was a very nice guy and later became a very devout Christian, but he was very prejudiced against anyone with long hair. One time on the way back to the office from a job where we were working in the mud, Chris was picking the mud off of his boots and it was clay-like and sticky, so he began to roll it into balls and throw it out the window at anyone with hair over their ears! I had to laugh at this crazy behavior as he kept saying, “There’s a hippy, watch this!” And then, *whack!* Someone would get hit right in the forehead with a clod of mud. I told him to stop and that he would get us both in trouble, but he would exclaim, “No, look, this guy is just begging for it; his hair is way too long!” Then he’d let another fly right at the guy’s chest! I was scared, angry, and laughing so hard I thought I'd wreck the truck!

Chris was funny in the way he would do goofy things that would make me mad but I would never get too mad because I could see that either he was doing it purposely to be funny or else he was ignorant and naïve enough to not know any better.

One example of this behavior was his apple science experiment. Chris left an apple to sit on his bookshelf in his cubicle until it was past rotten and began to dry. Chris said it was an experiment, or sometimes referred to it as a shrunken head. He gave strict orders to not touch it. I threatened to throw it away so many times that he hid it. I thought he had thrown it out and was glad of it. One day he came back from working in the field with me and said, "Oh, that's right, I have to check on my experiment." He came out with the apple, or what was left—a small vestige of an apple—and showed me the progress. His face broke into a wide toothy grin when I started huffing and puffing about the gross thing. He went back into the supply room with it. I followed him and looked everywhere for the thing so I could throw it away. It was nowhere to be found. It was months later that he revealed his hiding spot: above the ceiling tile!

Chris used to aggravate me by saying things that he knew would make me mad. Whenever the weather started to get cold in the autumn, I commented on how chilly it was. It was probably the first frost in November and the temperature was about twenty-five degrees. When I said that it was a bit nippy, he replied, "What are you gonna do when winter gets here?" This made me very defensive and I started exclaiming about how many years I worked in the field and how I spent days in the woods while hunting deer while it was freezing cold. He laughed at my sudden outburst, and then I realized that he said that simply to get my goat.

I also admitted one time I had tried drugs when I was younger and then gave my heart to God and was able to stay away from those evils. I expected some sympathy or perhaps some mild form of congratulations, but instead, he simply looked at me and said, "Don't bring that stuff to work." I explained with increasing vigor how that was the old me and I never even had any drugs anymore. He simply shook his head and said, "I'm tellin ya'—don't bring that stuff to work." When I finally got angry and practically shouted that I didn't do any drugs, he laughed. I eventually had to laugh too because I had never seen a guy that could get me so angry and eventually realize he was doing it for his own entertainment.

Since Chris was from the populated shore town of Somerset, New Jersey, he didn't know very much about cattle. I grew up next to my great-uncle's farm, where the Holstein cows would graze in the pasture. I knew the difference between a bull and a cow. There weren't any steers on that farm, but I knew that they were excitable, but about as harmless as a cow.

One day Chris and I were sent to do a topographic survey near Cross Keys, a village between New Oxford and Abbottstown, Pennsylvania. The Brethren Home was expanding their retirement village and they needed us to survey the area and make a topographic map for engineering design. When we got to the site, we realized it was a pasture with Black Angus steers in it. Chris immediately expressed his anxiety over going into that fenced pasture. I told him that we could walk into that area and I'd show him that the steers weren't hostile.

He kept insisting some of them were bulls. I ascertained that they were not and assured him we would not be charged. Anyone familiar with steers knows that they are a bit excitable and curious, so they followed us around and got close at times. When we moved toward them, they scampered about like puppies. Chris was still scared and protested against the idea of going into that field with the steers.

I went to the front desk at the retirement home and asked if the steers could be moved. They called the farmer and I spoke to him on the phone. He sounded a bit miffed that we were bothering him about this. I told him we were a bit worried about the cattle and he scoffed at me and said, "They're only steers; they won't hurt you." I explained that they got a little too close to comfort and he said to just wave our arms, or stomp the ground to disperse them and get them away from us. I thanked him and felt like we had our answer, so I told Chris not to worry; he had a twenty-five-foot fiberglass rod and he could use it to poke at them and keep them away.

This process worked as we took some shots along a stream but the steers kept getting closer and crowding around Chris. He got upset and eventually yelled and poked at them but they kept coming back. Then he got really mad and swung the rod over his head while running toward them to move them completely away from his area of work. Well, that started an almost hilarious stampede.

They ran at full speed up the hill toward a small loafing shed that was there for some shelter. The funniest part was the whole herd of about twenty head ran into the shed as I tried to understand how they were all fitting inside of this shed. As the last one entered the shed, the first one exited just as fast and the whole herd was coming back out at full speed! This made both of us laugh and so Chris ran at them again and started whooping to keep the whole stampede going!

As we stood there laughing we heard a truck pulling up to the side of the fence in the adjoining hayfield. The farmer got out of the truck and began yelling at us to get over there. We ran over to him and found out he was very angry.

He started yelling at Chris for scaring his steers and then laughing at them. Then he said, "You think it's funny to see them run around—you'd laugh if they broke through the fence and broke their necks, too." Then it was Chris' turn to get mad and he told the farmer, "You don't know me; you don't know what I would do." I pleaded for everyone to calm down. I was happy there was a fence between us and the farmer because he gripped the wire between the barbs and looked like he was trying to pull it down. Chris explained that he was scared of the cattle because he didn't grow up around them, and I reminded the farmer I had requested that he move them to another field and then he calmed down. We all apologized to each other and, in the end, I thought he was going to invite us over for dinner!

Chris used to pack his lunch each day and it usually consisted of a big Kaiser roll that his wife got from the bakery within the farmer's market where she worked. He usually stuffed it with turkey or ham and then at lunchtime he would eat so fast that he nearly inhaled it. I would watch him from the corner of my eye, while I enjoyed my sandwich, eating like he was a starving man.

He usually asked to go to the Turkey Hill Mini-Market for some of their diet sweet tea. Most days, he had the sandwich finished before we even got to the store. He gulped and chewed and swallowed with such fervor that I couldn't help but ask, "Are you hungry?" He would just nod emphatically while gulping down some more. Then he invariably would groan, holding his belly and start to hiccup non-stop and moan about how much it hurt to hiccup so violently. He would say, "Oh no, I ate too fast," as he would clutch his stomach and his throat.

The next day as he started to eat like a hungry wolf; I would say, "You know you're going to have a sore belly and a sore throat from hiccupping if you eat that fast", and he would yell, "I don't care, I'm too hungry." He always said these terse comments about his hunger, his eventual pain, and incessant hiccupping like he was scolding a child or had to repeat instructions when he was out of patience. The same routine went on just about every day. The only exception was when his wife brought some leftover "cow pies" from the market. They were actually glazed apple fritters that were delicious with a cup of coffee, but Chris insisted on calling them cow pies because of their piled-up appearance.

Chris loved to harass this one guy who worked in our office, named Thom Rudloff, about his hair and his name.

The Long-hair

Thom Rudloff was the one referenced above with shoulder-length hair. It was what I would call strawberry-blonde, but Thom hated that description. He was a stylish metropolitan type, which was ironic because he came from the coal-region town of Tamaqua. He listened to a jazz-fusion band by the name of Weather Report and wore the fashionable clothes of the day.

He also befriended me by inviting me to his apartment after work to watch the *Saturday Night Live* shows that he had recorded on VHS tapes. He knew that I enjoyed comedy and couldn't stay up on Saturday nights to watch the show live—at one point, we didn't even have a TV—so we would go over to his place on Monday after work and laugh at the skits with Joe Piscopo, Billy Crystal, and Eddie Murphy.

We loved the sketch in which Billy Crystal played a security guard. The characters were Willie and Frankie and they constantly talked about painful ordeals and then ended each crazy description of these with, "I hate when that happens." We practiced the lines and repeated them in the office to get some laughs. I'd say, "The other day I was sittin' around the house looking at a ... uh ..." and then Thom would interject, "Scale model replica of the Empire State Building?" and I would say, "Yeah, yeah. Then I tried to shove it up my nose; to see if it would fit, so I hammered it in with a ... uh, uh ..." and Thom would say, "A 12-ounce ball-peen hammer," to which I'd quickly agree, "Yeah, yeah; gee talk about hurt!" Thom would finish the routine by saying, "Gee, I hate when that happens."

One of the funny things Thom had said was about how he got the mounted sailfish that he had hung over his desk. The fish appeared there after we moved into our new offices at 3020 Columbia Avenue and I believe the mount was a trophy from a fishing trip that Hank had taken. I guess he didn't want it in his office and since it was up for grabs, Thom thought it would look nice at his desk. Thom called it the Gefilte Fish, and said that he caught it on a fishing trip in the ocean. I didn't get it. Then one day in a store with kosher food I saw a jar of gefilte fish. It's a kosher food that Jewish people like to eat made of carp and other fish flesh. I realized later that Groucho Marx had made the same joke in one of the Marx Brothers movies; calling a live fish a gefilte fish.

The other laughs Thom and I shared were about a story he told me involving his friend that always called him, Butch. I thought it was amusing, so Thom started calling me Butch. Not really funny, but I always said he looked like a girl with long hair, so he was sort of like calling me a butch-looking guy. The meanest and yet funniest thing I did to Thom was on my last day at Rettew. I paged him on the phone—this was after three years and the company had grown to about sixty people—and said, "Thom, your hairdresser, Raul, is on line one!" He chased me through the whole survey department, trying to beat me up. I had to beg for mercy while trying to catch my breath from laughing so hard.

The Ole Cockbird

Another one of the guys that Chris Embry liked to pick on was a fellow by the name of Fred. He came to Rettew shortly after finishing his time in the Army. Fred thought he knew everything about surveying by what he was taught in the Army. He was a little larger around the waist than the rest of us, so Chris labeled him "Chitlin!" This name was very funny to us, and the name stuck, much to Fred's chagrin.

Fred was known to stretch the truth once or twice and we loved to get him started about hunting stories so that we could listen to some "whoppers." One of his stories was about his father going to hunt pheasants one day and he saw a flock of "tame" pen-raised birds that had recently been released. According to the story, the birds heard the hunter leaving the truck and they ran straight toward him. Fred told us his dad turned, got in the truck and left, and, "... he hasn't hunted since then." Chris had a funny comeback—he asked, "Shook him up pretty bad, huh?" As if pheasants running toward him would be too traumatic.

From that day on Chris had a new name for Fred—it was "the Cockbird." Fred had an old yellow Chrysler with a big V-8 engine. Fred would go outside and put on a "smoke show," by burning rubber on the pavement anytime we goaded him into it. Then the car became known as the "Cockbirdmobile."

One day while Chris and Fred were working with me, we took our mid-day break and sat in the truck to eat our packed lunch. We always worked out of a pick-up truck with a cap to cover the equipment. It was fine for a two-man crew, but with three in the one seat, it was kind of crowded.

We were eating our lunch when I noticed Fred had a big bag of potato chips he was digging into. He explained that it was only half full so he decided to bring the whole bag along to work. As he was eating big handfuls of chips over on the passenger side, I noticed Chris leaning into me from his position in the middle of the seat, as if he might fall into the huge mouth where these handfuls of chips were going. I started to laugh at that one without Fred realizing what was so funny.

Then he produced this two-liter bottle of soda—some kind of red-colored, cherry-flavored stuff—and began chugging that to wash down the chips. Chris and I were just about howling at this display of gluttony when suddenly, Fred let loose with a loud belch. Now Fred loved to belch loudly with the sound of DAP. We always laughed at him for this belching sound. It sounded just like the brand name of caulking, but this time it sounded more like DrrAP! Some of the red soda foamed out over his lip and down over his chin. Fred tried to catch it and wipe it off before we could see it. Well, that made it even funnier, since we were already amazed at how sloppy and gluttonous he was being. I really didn't want to make fun of his eating habits, but I didn't know a better way to point out how uncomfortable it was to share lunch in the close quarters of a truck cab with someone who was so raucous in his style of eating.

At Rettew there was an old Chevy Suburban (1971 model) that we called the "Perc Truck." This name was derived from the use of it to haul water to job sites where percolation tests for on-lot sewage disposal suitability were being performed. I never saw it used for anything except hauling wooden stakes from a sawmill in Leola to our office in Lancaster. Usually, on rain days, someone would get assigned the task of driving out to Leola to get stakes. One day when Chris was going out to get stakes, he came back in and asked me to help him jump-start the perc truck because the battery was dead. After I connected the jumper cables, I waited for him to start the motor. As the starter engine turned the motor over and over, it wouldn't start. I glanced into the engine compartment and saw a spark at the distributor cap. I was no mechanic but I knew that meant there were electrical problems and the engine was not going to start. We unloaded my company-owned survey truck and used it to get stakes.

Hank Rettew wanted to get that useless truck out of the parking lot and offered it for sale. There was no interest. I decided to buy a new rotor, distributor cap, and coil wire, based on some advice by mechanically inclined friends, and install them to try to get the truck started. When I connected the jumper cables and hit the ignition the engine fired and started running! I had to play around with the shifter by starting it in gear and jerking it out of gear to free up the clutch, but soon I was driving it around the parking lot. I went in to the office and asked Hank how much he wanted for the truck. He said, "If you can get it off of the lot, I'll sign the title over to you." I had a truck for the price of a few ignition parts! With the help of some friends, I put aluminum panels in the floorboards and some body putty on the fenders and was able to get it inspected and licensed for the road.

The best thing about Rettew was the number of Christians that worked in the survey department. We all got along well until the number of non-believers grew to over three. One of the guys even complained that I was "preaching" too much. I was surprised! I was always afraid that I didn't share my beliefs enough. At one point there were nine guys in the department and eight of them were Christians.

Ed the Evangelist

Ed Warfel was probably the most evangelical guy that I ever worked with. I'm not going to debate whether his methods were done the correct way or even if religious beliefs should be discussed at work, but I had to admire him for living his faith every day. It was a little bit uncomfortable to watch him share his faith, though.

When a new guy would come on the crew, on a three-man crew, and we'd be driving to the job site or returning to the office, Ed would look at the new guy and ask about where he went to church. If it wasn't an evangelical church, Ed would ask what he believed about heaven. The conversation would eventually lead to Ed asking, "If you were to die in an accident on your way home tonight—and I hope and pray this doesn't happen—you would immediately be standing before God. If God asked you why you should be allowed in heaven, what would you say?"

Well, now the new guy and I would both be feeling uncomfortable. I was just feeling sympathy for the guy who really didn't want to be put in that spot, but Ed was doing this out of compassion; he really wanted to lay out God's plan for salvation. I couldn't fault him for that because his motives were pure.

Ed and I also got into some religious debates because he was a Calvinist, one who believes that you are chosen by God to be saved and that once you are saved, you are always saved. It used to bother me a lot and we spent hours while driving debating this, and then one day I heard a Christian minister say it is a non-issue; if you're saved, only you and God know it. After all, it's only God that really matters.

Big Norm

One of the guys that was a bit different from the rest was a big fellow named Norm. This guy weighed about 300 pounds. The TV show, *Cheers,* was popular then and we made the obvious connection with the fat guy on the show by the same name.

Norm didn't wear a coat in the spring when he started working at Rettew. He said coats made him sweat and then when the cold air would hit him, he would get cold and sick. Even when winter came, Norm would wear the same blue sweatshirt, with gloves and a hat if the temperature was below twenty-five degrees! I don't know if it was the fact that he wore the same shirt every day or if he had other hygiene problems, but Norm had a body odor about him that smelled like dog droppings!

I could hardly bear sitting in the same truck with him. He was also a good liar. Some of the guys who worked with him regularly began relating tales of him going places during the day that were not work-related. One time he left a job site in Columbia, Pennsylvania, and drove, with the instrument operator along, to his house in Reading—about fifty miles away! He explained that a contractor was doing concrete work for him and he had to check on the guy. Another time, he made a similar trip with another guy and then took a shopping trip to the outlet mall. He eventually got turned in, confessed, and then denied realizing any problem with that type of action.

The part of his lying ways that affected me was how he would request a third guy for his crew because he was doing stakeout and could use some help. Well, I was doing stakeout, too, but I was expected to get by with just two guys—me and the instrument man.

Then, I found out that when he had three guys on the crew, he would sit in the truck and let the other two do all of the work, while he pretended to study the plans and call out the stationing to be written on the stakes! I was appalled by his actions and was not surprised years later when I heard he was fired for dishonesty.

A Surveyor from the Old School

An experienced—read "old"—guy that started working there in the survey department was a licensed surveyor named Gene Lutz—"Rhymes with *nuts* which is where I'll kick you if you say it *Loots*—" That is what Gene said when asked how to pronounce his name. He was probably about fifty when he came to work there and most of us were in our twenties. Even the bosses were in their thirties, so Gene really brought some experience to the place. He told us stories of his early days working in the City of Philadelphia. They didn't have a survey truck; they just walked to the job or rode a streetcar if it was across town. Can you imagine jumping on a streetcar with the transit, tape, rod, and other assorted tools in hand? I guess they rarely needed a machete or brush hook, but that was definitely a different time and place than when I surveyed. I have to admit, I picked on Gene because he was old enough to be my father but he took it all good-naturedly. He did teach me some of the finer points of boundary resolution.

When I got a little bit too repugnant for his liking, he would look at me sternly and say, "Go to your room." It was mostly in fun, so I took it in stride.

Mr. Mathematics

Geoff Kurtz was our resident computer expert for a while whenever we first switched from the old tape-driven HP computers to the IBM XT computers. I remember the first full-color monitors that could make a pie chart! That was some cool technology back then.

Geoff had a degree in mathematics and started his career working in a bank; after regular banking hours, he entered data into computers. Then he got a job at an engineering firm, named Huth. There again, he worked with the old-style computers, so he was a pioneer in the computing world that I knew back then.

He became a licensed surveyor, although I think his field time was limited. He taught me a lot about computers and how to compute coordinates. I believe Geoff was happy to be a technician, but he was eventually pushed into more of a project management role. When he became a project manager (mostly during my second term at Rettew, 1995-1999) and he had a project assigned to him, he would often respond to inquiries with, "I know next to nothing about that!" The last word would come out as, "thaa-at." We would often laugh when we were told to see Geoff about a certain project because we knew what to expect.

I worked at Rettew for three years and we had our second son named Shane. We wanted to build a house, so I started looking for a new job near our parents where we could afford to build. When I had taken the job, we first moved into an apartment in Red Lion and the first night I heard so much noise of cars trying to get up a hill in the snow, with wheels spinning and doors slamming, that I wondered if I would be content there.

Later we got new upstairs neighbors that played loud music and fought so loudly that we had to call the cops. We also soon determined that they were selling drugs and had a lot of questionable characters coming and going at all hours. To have better living conditions, we traded the AMC Eagle as a down payment on a mobile home and moved it into a trailer park near East Prospect. There were some good neighbors there, but there was too much noise and the neighbors were still too close! This was where we lived when Shane was born and two other women that lived near to us, had babies within two weeks of his birth.

I still had friends at Fox & Associates that tried to convince me to come back. I talked to the new owners who were some of my old bosses and they convinced me that "things are different and better," so I returned.

Part V

Fox & Associates, Inc. (second time)
July 1987 – January 1991

When I first started this new job, I had to live with my parents for a few weeks while Trudy and the boys stayed in East Prospect in the mobile home. We visited on weekends and searched for a place to move our trailer. We finally found a place in Mont Alto, just down the road from the entrance to the Penn State campus. It was owned by a lady named Mrs. Williams. The rent was decent, the setting was nice with a large backyard, even though the road in front of the lot was a busy state highway. We found a man who moved trailers for a cheap rate, but we had to provide the escort vehicle with an amber light and placards that notified drivers of the "Wide Load." We also had to move the concrete blocks that were used as the foundation and the skirting that trimmed out the bottom of the trailer. My dad agreed to haul the blocks and skirting and I planned to drive the escort vehicle. I used the old Chevy Suburban "Perc" truck that I got from Hank Rettew.

The night before the big move, while I was sleeping at my parent's home, I got a stomach virus and started to vomit and have diarrhea throughout the night. I barely slept. In the morning, I couldn't eat and barely drank any water.

We already had so many arrangements that day to make this move that I couldn't just lay around.

I took off driving with Dad behind me in his Jeep and utility trailer. I told him to expect me to pull over to throw up whatever water I drank. Somehow, we made that trip without any stops for bathroom use or puking! When we got the trailer to the lot in Mont Alto, I went inside and lay on the floor while the driver jacked up the trailer, set the blocks under it, and got it level. If I hadn't been a fit 26-year-old, I don't think I could have accomplished that.

Mrs. Williams came out of her house to see what the trailer

looked like, and all that she said was, "That looks big." The next day she called me to say that our trailer was longer than 70 feet and that there wasn't enough room to get another trailer in next to ours. When I assured her that no trailers were being manufactured that were longer than 70 feet, she demanded that we move the trailer forward. There was no way that would be possible and I told her so. This started a rocky relationship with her and we couldn't wait until we could build a house on some property that Trudy's parents were willing to give us.

Some of the same guys at Fox & Associates were still there from my first engagement with this firm. Russ was still there as the chief of surveys. Tim Witter was now working in the office as a licensed surveyor doing the scheduling and my old party chief, Mike, was still a party chief. I found out later that for the company to pay me what I wanted for wages and still have Mike make more than me, they had to give Mike a raise. That didn't do much for our friendship because Mike was upset that I could make as much as he had been making for staying there all those years.

When I reported to Tim's office for work on the first day, the other party chiefs were coming in to get their assignments as well. One older guy came in and asked, "Who's going with me today?" Tim had to tell him, "Well, Dave, you're going with Eric today." Sort of a demotion on the spot! That one sentence told him he was no longer a party chief and, worse than that, he was working with the new guy. I don't think that made him fond of me. We worked together for a few months and got along alright, but there was always some tension. One thing he would often say—starting with the first day—was, "Now a good party chief will always ..." and then he'd fill in the blank with whatever it was I wasn't doing that he wanted me to do, such as buying coffee for the crew, letting everyone go home early with pay, or taking an afternoon nap! I really worked hard because I wanted to prove myself.

Dave had been a truck driver before his foray into surveying. He once told me that he was a really good truck driver; he just couldn't back up and turn around very well! I laughed incredulously at that and said, "Just about anyone can learn to shift gears and drive ahead down the road: the real skill is in backing up and turning around.

Florida Man

I was soon assigned a new instrument operator to work with, Ed Fletcher. Ed was a Vietnam veteran. He had been wounded and was receiving disability payments from the VA, but that didn't stop him from working forty hours a week; he just couldn't carry much weight.

His injury was to the back of his arm so he claimed general weakness in his upper body. Ed was originally from Florida so he spoke with the southern drawl and had some funny sayings. One thing he said when speaking of the past was "back in a day." I had never heard that one, so I had to laugh. Ed was a good liar. He liked to tell stories that made him look like a hero, but no one would believe him.

One of Ed's biggest whoppers was one that he told me when I had bought a new deer rifle with the money I had gained from selling the old Perc truck. We began talking about which caliber was the best. I had bought a .30-06 Springfield rifle and Ed tried to tell me how much better a .35 Remington was. I argued that a .30-06 was much faster and had a flatter bullet trajectory.

Ed told me about a .35 caliber rifle that his friend bought and they took it to a junkyard to shoot it. It had open sights but they set up a license plate with the numeral "0" on it and paced off three hundred yards. Ed shot it and hit the "0" dead center, his friend shot right beside it! I gave Ed several chances to change the story, but he stuck with it; so, I called him a liar to his face! There's just no way to sight that far above the target (the .35 drops about one hundred inches in that range) to even hit near a license plate.

Ed would always try to impress me with his demonstrations of surveying prowess. He claimed loss of use of his left arm, due to his war injury, but he would sometimes volunteer to pound a hub in using a sledgehammer one-handed. He liked to chew sunflower seeds incessantly—he bought them in bulk—and he would consume about one pound each day!

As we would look at the plans while he crunched, spat, and chewed, he would talk loudly about what he thought he knew, with his mouth half-full of un-cracked seeds. If we were doing reconnaissance of property corners and pulling the prescribed distance for the next corner, he would rush ahead as soon as he could anticipate where the measured distance would fall and start furiously kicking the weeds down and if he had the magnetic locater, he'd start waving it around under every bush. If he found anything plastic with some color—like an old balloon—or if he got any reading with the magnetic locator, he'd yell, "I got it! I knew it was right here." When I'd ask what he found, he'd say, "I didn't see it yet, but it's right here." Eventually, he'd realize that the corner wasn't there and sometimes the distance would be thirty or more feet past where he had been looking.

Ed used to complain about another guy that worked as an instrument operator on one of the other crews who was also a disabled veteran of the Vietnam War. Ed said that this fellow, Jim Higgins, was one of those vets that thought the world owed him something. The funny thing was that Jim said the same thing about Ed. I didn't know Jim very well until he was assigned to my crew. I knew that he wore out his welcome on Pat's crew and he was working with another party chief named Tony. One day Tony came into the chief of surveys' office early and shut the door. He said to Tim Witter, "Look, I want to say something before the instrument operators come in here: I want that Jim Higgins off of my crew!"

That afternoon I was told I was the chosen one to receive this guy on my crew and he had been warned if anything went wrong, he would be fired. I was told to watch him like a hawk and report any wrongdoings. Well, I always was an old softie, so I began treating him with respect and kindness—thinking that he would respond with good behavior and hard work. I was naïve enough to think that everyone had the same kind of values and morals with which I was raised.

Nice Guys Finish Last

This Jim Higgins was quite a piece of work. He was about thirty-eight years old at the time; he had no car, no house, rented a small apartment, and also rented all of the furnishings. I started my project of turning him into a "good" person by picking him up for work every day. This was after I watched him start walking toward his apartment through the rain after we were discharged for the day because of the inclement weather. I couldn't stand the thought of someone walking for 45 minutes in a cold drizzle, so I drove him home and offered to pick him up the next morning.

When he spent all of his money on booze, pot, and renting a large screen TV, I would give him some of my lunch instead of watching him drink a coffee and smoke a cigarette. That year at Christmas, I invited him to my house for dinner. Trudy cooked some deer steaks and all the trimmings. I drove to Hagerstown to pick him up, drove him to our mobile home in Mont Alto, and then drove him back home again. We even sent some frozen deer meat home with him as a present. He repaid me by coming to work hungover, barely able to do his work, smoking pot on the job (I learned this later from some of the guys that smoked with him), and lying to me about almost everything. The rest of the party chiefs were pretty mad at me for not getting him fired. They even resented my driving him to work! I don't regret all of the things that I did for him because I felt I was displaying the love of Christ, but now that I'm older, wiser, and maybe even cynical, I probably wouldn't have gone so far for him.

The Worst Boss, Ever

The "big boss" in the surveying department was Russ. I don't know exactly what his title was, but he was supposedly a partner or part owner or something in the firm of Fox & Associates, Inc. He promised me "things are different" and "this is a new company." Well, there were some differences—the main one being that now Russ, Bill Weikert, and Fred Papa were the bosses and Bill Fox was mostly a silent partner. So was Bill's old partner: Bernie Charles; in fact, it wasn't long after I rejoined the firm that Bernie died. I guess it was around 1989. I remember we got the afternoon off with pay, but I didn't attend the funeral.

However, one thing that didn't change—Russ was still an old grouch. He treated clients like gold and I guess he was real cordial to the other partners too, but he treated his underlings like dirt under his fingernails. He was not very personable and most people regarded him as an oaf. Russ was about six feet and four inches tall and weighed well over 300 pounds.

Just imagine that hulk plodding about the office with a huge beard and little, cherry-like lips that usually were formed in the shape of a carp mouth. He rarely smiled except when someone was being intimidated or embarrassed. Russ liked to look at his watch as the party chief would walk into his office at the end of the day. He always wore it on the inside of his wrist and he'd flip his wrist around to look at the watch several times before asking, "Done yet?" If we answered with a negative, he'd shoot back, "Why are you back?" Even if we told him the project that we were working on would take another full day or so, he would expect us to work late to get more done! I guess he never realized that we had personal lives with wives, kids, and things to do.

Russ didn't like anyone to have fun, probably not even himself. When I worked in the office, I sat in the corner of the drafting room where there was a computer loaded with a coordinate geometry program. We didn't have any computer-aided drafting (CAD) back then; the room was full of drafting tables (about six of them) with drafters busy doing all types of plans and plats. The two tables in front of me were the workstations for Steve and Martha, who did the house location (mortgage) surveys. They did the courthouse research, deed plots, fieldwork, and then the survey drawings. When they had questions, they always asked me for help.

I had more experience and could often help them to complete their work. They also reported to Russ because he usually sealed the drawings as the registered surveyor. Steve was a young, fun-loving single guy with long hair. He loved some of the same rock and roll that I was excited about and he told us stories of his "life on the road" with Twisted Sister and some others as a roadie. That was fun to hear about, so we had a good friendship.

The phones in the office could be set to play the "hold" music—that callers would hear when they were put on hold—from the local radio station, WQCM, which played pop music. Steve and I would have fun singing along and commenting on the songs that were popular back then. I always think of those days when I hear Edie Brickell and the New Bohemians, singing their song, "What I Am" (Shove me in to shallow waters ...).

Martha was a small young lady. She was about five feet-nothing and must have weighed about ninety-five pounds soaking wet. She was a really nice person, with a sparkling personality and a southern drawl that made her sound like she was from Texas. She was actually from Williamsport, Maryland. I noticed this about folks from rural Maryland; they had southern accents as if they were from southern Virginia. I found this to be true of folks from Williamsport and Walkersville (near Frederick). Martha was an engineering graduate from the University of Maryland. She didn't want to pursue a career in engineering; she just enjoyed her work and the social life after work similar to what she had in college.

There was one particular survey that Steve and Martha were working on which needed some input or a decision from Russ. After they discussed the problem with me, I asked Russ to come to join us and look at the drawings, deeds, etc. As we were reviewing the documents, Steve swung his arm and spilled his Big Gulp fountain drink from 7-Eleven. The soda came cascading down the top of the drafting table (set at a fifteen-degree slope, as they usually are) directly onto my pants, just below my waist.

As I recoiled in disgust and anger, Steve kind of chuckled and Russ began to laugh as I had never seen him do before. His round belly bounced up and down and his eyes squinted to almost closed as his mouth broke into a wide laughing smile. He kept trying to make everyone else laugh harder by squeaking out disparaging remarks between his grunted laughter. He pointed at my crotch and said, "What happened?" And then, "Do you need big-boy Pampers?"

Then he looked around and said, "Someone get some paper towels, Eric had an accident." All six of the drafters in the room joined in because they were either amused or just thought that they should laugh if the boss was laughing. I didn't see the humor in it as I realized that I would spend the remaining two or three hours of the day in wet, sticky pants. Russ even admonished me that I shouldn't have left my drink setting on the drafting table and my protests and explanations that it was Steve's were muffled by the laughter in the room.

This was one of the few times that I saw Russ laughing, albeit, at my expense!

As I worked with Steve and Martha every day, we talked and became more and more familiar and would sometimes allow our work conversations to drift into personal conversations and jokes. I always had the philosophy that *we have to spend over eight hours together at work; we may as well have some fun while we do our work*, but that was not the philosophy of Russ.

I noticed the drafters in the room were mostly quiet while working. This was a big room without any cubicle dividers; the drafting tables were almost touching, so everyone was close to one another, and eventually, quiet conversations would start. When Russ entered the room, apparently because he could hear the murmurs down the hallway, everyone would put their head down in a position of earnest tracing, lettering, or scaling on their mylar drawing sheets. Then Russ would walk back the hall to his office, content that he was getting the most out of his underlings.

Well, as I said, the conversations with Steve and Martha would sometimes get animated. I love telling jokes and funny stories, especially when people are laughing along with me. Neither of my workstation neighbors knew how to keep quiet when they were having fun—both of them laughed rather loudly. When they got wound up, they also started talking louder, as if they learned to whisper in a sawmill!

As I said, I was in the corner of the room, so I had a view of the hallway where it entered the drafting room. When I saw Russ enter, I would shut up and that signaled Steve and Martha to do the same. After a while, I got tired of the "cat and mouse" game so I would just continue with the conversation and wait to see what Russ would do. He stood there with his hands on his hips and said, "Don't you have enough work?" After about the third time of this happening in a week, he finally said, "You guys need to keep your mouths shut and get your work done." Everyone in the room had their heads tilted forward and their pens moving furiously except for me. I asked how he was judging the amount of work being done because we were all keeping up with the workload and producing a lot of drawings. He just said, "You can get a lot more done if you're not talking!"

Well, this didn't sit well with me because I was used to working in the field and usually with some fun-loving people. Shortly after this event, I happened to be in the Washington County Courthouse Land Records office, in downtown Hagerstown, and I noticed a funny little sign that someone had posted. It was a yellow smiley face (you know, the one that was supposedly created by Forrest Gump, in the movie by the same name), but this smiley face had a red prohibited symbol (red circle with a slash line through it) across the face. Below were the words, "No Fun Allowed." I told my counterparts at the office about this. They thought it was funny and remarked how we should have that sign in our corner of the office. That was such a splendid idea that I got at it and made one in full color using colored pencils. This was well before the word processor or anything like that so there was my hand-printing for all to enjoy. I think maybe Jim or Kim, two of the drafters who shared the room, may have told me that I would get in trouble, but that just made it more of a dare and something that I was proud of.

Then one day it was gone! I assumed that Russ just tore it down and threw it away, but then he called me into his office and said, "Close the door." As I returned to his desk from closing the door, he pulled out a wrinkled piece of paper and spun it as it slid across the desk toward me. It was my "No Fun Allowed" poster. He gave me a grim stare and questioned, "Is that your doin's?" I nodded and he asked, "What's that supposed to mean?" I told him why I put it there, and how everyone was afraid to talk or have any fun at work and then finished by saying it was just put there to have fun. He muttered, "If you want to have fun, you can go work somewhere else!" I simply said, "Okay" and walked away. I had to decide then if I should just keep walking out into the parking lot and go home, but I felt that I should stick around and finish the day and just keep putting up with that kind of work environment because I really didn't have any other choices. I needed the job and they knew it. I was trying to build a house and needed a steady income.

The Foreigner

One of the drafters that worked with us in those days was a lady from Poland who I believe was there under a work visa. Her husband was a mechanical engineer who worked at Grove Cranes. Her name was Grazina (we called her Gina) Gulinska. She had a very thick accent and a good level of conversational English, but some engineering terms and specific names of items stumped her. I was one of the few who would listen to her and could understand her accent so she would often ask me for help in understanding terms from the plans or field notes. She asked me if a silt fence was just like any other fence made for keeping animals in or people out. I had to explain that it was just a fabric that would allow water through and silt out. She also wondered about the ornamental shrub, yew. I spent a while trying to explain how it sounds like you or ewe but it was completely different.

At that time most people with a normal income owned a video cassette recorder/player (VCR). I wasn't sure how well Poland had kept up with modern household technology, so I asked her if they had them in their homes. She looked at me with some disdain and said, "Yes, we have them, we also have TVs, we have refrigerators!" It was pretty clear that I had offended her and I apologized, but she still acted a bit miffed. I didn't try to make many conversations with her after that.

The management at the company held a meeting one afternoon to announce that our health insurance was once again going up in cost. None of us were happy about this, but we knew there was nothing to do but pay for it if we wanted insurance. Gina began to complain loudly and said, "I don't know why we have to pay for this in this country; in Poland, the government pays for all of that. If you go to the doctor, there's no cost, if you go to the hospital and get an operation, there's no charge."

Now, I had never really spoken to someone from a country with socialized medicine, but I kind of figured out that the government can't give anything to the citizens that they haven't already taken from the citizens. I asked her a few questions to clarify what she was telling us and then asked her, "What percentage of your paycheck goes to the government for taxes in Poland?" She told me it wasn't that much. When I pressed her for an estimate, she admitted that it was almost fifty percent of the paycheck! I asked her if it wasn't better to have an opportunity to make more money here and manage your own money, paying for insurance if you wanted it (this was before the Affordable Care Act—Obamacare) or saving up money for healthcare if you liked. She said, "No, I liked it better to have the government take care of it." I responded with the only retort I could imagine at that moment and said, "Well, why don't you just go back there then?" She got silent. The cold shoulder wouldn't adequately describe the chill in the air. I simply finished the conversation by reminding her that she was there because she and her husband were enjoying a comfortable life in the United States and said, "Admit it; you're a capitalist or you wouldn't be here!" We didn't talk a lot after that.

Company Policy

At a certain point, the company decided to make some new policies to improve morale. They wanted everyone to join them after work for beer and snacks at a hotel conference room and discuss the company policies. This would have been a good idea if they were open to discussion and ideas, but I knew better. They kept telling us that we really needed to come to this meeting to voice our opinions and share our views. I told them I didn't think they really wanted to hear my opinions. All of the management said they wanted everyone to come and speak up, that they wouldn't make any decisions that night, but instead, they would listen and make decisions later.

Well, that didn't happen! As soon as something was brought up for discussion that would cost the company some money, they would say, "Oh no, we can't do that." This included my idea for being paid for travel time.

The policies definitely needed change, but they were only interested if it didn't cost anything out of their profit-sharing bonuses. Travel time—when the survey crew left the office (usually 7:00 a.m.) the clock started for their pay. The clock stopped when they left the job site. So, a typical day would start at 7:00 a.m. and we would take a half-hour lunch, so, we had our 8 hours in at 3:30 pm. Depending on where we worked, we might not get back until after 4:00, but that travel back to the office was on our own time.

That was one thing that made me mad. If we were working in Rockville or Gaithersburg, we might get back at 5:00 p.m. We were forced to drive up to an hour on our own time!

The company also forced us to work the day before and the day after a holiday, with only one exception—if you had scheduled vacation on the day before or after the holiday.

One year I went on a day trip with my wife and kids on Memorial Day to Potter County, Pennsylvania. We drove up there early in the morning and by afternoon, I was sick to almost death. I couldn't drive home, my throat hurt, my head hurt and I just slept.

The next day I couldn't even crawl out of bed. It was the only time in my career that I had my wife call into work to say I couldn't make it. I went to the doctor that day and found I had strep throat. When I went to management, first to Russ and then the others, but they refused to make an exception, even after I produced a doctor's excuse and the lab results of a throat culture which had the word, "Streptococcus" on it. That made no difference to them. "We have to abide by the company policy; no exceptions."

I reminded them of people who left for Ocean City on Thursday night, called in sick on Friday, and got a three-day weekend. That made no difference. I told them that I would get my eight hours of pay one way or another. I said I would be sitting on the job and collecting pay on days when I just didn't feel like working until I had accumulated eight hours of pay for not working. Russ got really pissed off and told me that I'd be fired if he caught me sitting in the truck and not working. I simply replied that he wouldn't catch me, I was too smart for that.

It was not very conscientious of me, especially when one considers that there were two other members of the crew who were sitting in the truck and not working as well. I just felt rebellious enough to do it so there were days when it was cold or hot or I was just in a mood to not work, that I would sit in the truck and read the paper or we would just go for a drive.

One of the other policies that frustrated me was the use of company vehicles—the party chiefs were allowed (actually directed) to drive the work trucks home. After I became licensed in Maryland as a property line surveyor, I was brought inside to work in the office while someone else was working in the field with my crew and the truck I had been driving. Russ tried to explain to me how I was being promoted and groomed for the chief of surveys position in the Frederick office, but yet I had a major benefit taken from me.

Even when I tried to explain my position in needing to buy a vehicle for my use and how that was a real disadvantage to me, Russ said, "I don't understand." I tried to reason with him that forcing me to come in the office was actually punishing me and I didn't care if he agreed, but I asked if he could understand my point. He simply replied, "No."

That type of intellectual dishonesty continues to make me angry when I encounter it. We may not always agree, but I think you owe it to me to be intellectually honest; in other words, you have to admit if a person's opinion has merit and is logical, even if you don't agree with it. There's no way to have a discussion with someone who just refuses to see your point.

A Liar, a Cheat, and a Thief

There was one guy who worked on my crew by the name of Mike Davenport. He got laid off when things got slow and just before I transferred to the Frederick office. He was dating Martha Hildebrand—the girl who did the house location (mortgage) surveys, mentioned earlier—so he would come back and hang around the parking lot at quitting time. The odd thing was that he began driving a new car that he had just bought, even though he didn't have a job. We commented that it was a bit odd for someone who finds himself unemployed to take out a loan for a new car. Once when he asked one of the instrument men, Tim, "When are you going to buy a new car?" Tim said, "I guess I'll wait until I'm laid off like you did!"

Mike was known to stretch the truth a couple of times. He invited my family and me to swim in the pool at his parents' farm. When we got there, the pool was so cold that we couldn't even stay in it, but he offered to show the kids the tractor. Well, he had talked about this big enclosed cab John Deere, but when we went to the barn, there was a little old run-down Massey-Ferguson. When I asked about the big, new tractor, he said that it was at the neighbors. He had spun various other yarns about big deer that he had killed and had the heads mounted. When I asked to see them, he sputtered and then quickly came up with the excuse that they were kept at the hunting cabin.

Then one day when I was working in the Frederick office, I started to realize how far his deceit could go. Russ approached me about a bill that I had signed for from an office supply store. I thought at first that it was a work order for a copier repair, but Russ dismissed that suggestion and said, "No, this is for a $1,500 typewriter!" I had no idea what he was talking about. He showed me the invoice that had my name written in a script, but it was definitely not my signature! I pleaded innocent of all charges and pointed out that the signature was not even written with the same characters as I *use* for my signature. He said that they would be turning the matter over to the police.

I was contacted by the Maryland State Police to come to their barracks to have some evidence gathered against the accused thief. They took mug-shot-like pictures (front and profile), asked a lot of questions, and requested a lot of handwriting samples of my name and others with the same last name. Then they asked if I knew this Mike Davenport. When I replied in the positive, they said, "Don't take any checks from him," He was the one accused of the theft!

He had met a young girl and was bragging about how much money he had and then had to steal and pass bad checks to try to prove it. I was so upset that he would use my name to try and steal a typewriter—like I had that kind of authority within the firm and the preposterous notion that no one would question it! I was subpoenaed to appear in court, but the day before the court date, I was notified that he pleaded guilty and made restitution. Oh well, I learned that some people just can't be trusted or believed.

The Scholarly Instrument Man

After I was brought into the office and others were working in the field, the amount of surveying work diminished as the US was preparing for the Gulf War. A lot of the "dead wood" was laid off. The work became so scarce that they no longer needed me to work in the Frederick office and I was sent out in the field once again to work with a different instrument operator.

This was another guy that the other party chiefs didn't want on their crews. His name was Tim. He had been hired full-time after working on the crew as a temporary helper from Manpower (a temporary work agency). Tim had taken some classes at Hagerstown Junior College and proved to be intelligent, but his main fault, according to his detractors, was that he was not aggressive!

I liked Tim; he was intelligent and I felt his biggest problem was that he was not trained properly. His previous party chief had said that Tim couldn't read the rod accurately and questioned his eyesight when they had run a level loop and it didn't "close." The level loop is a way of determining elevations and when the loop is "closed" by returning to the starting point, the accuracy of the elevations can be determined. Apparently, their level loop had some errors. After I questioned Tim about how the level loop was set up, he revealed to me that their back sights and fore sights were not balanced, which is a big-time no-no. That is most likely what led to the errors in the elevations, and the errors were not a result of bad eyesight but rather bad logistical setup by the party chief.

I realized that Tim was more intelligent than the average instrument operator when he showed me the programs he had written for his Texas Instrument programmable calculator.

He quickly learned how to calculate curve deflection angles and cut sheets for grade stakes. The most fun we had was discussing the wacky and burnt-out hippies and dope heads we had known in our youth. He was a reformed pothead that couldn't quit smoking cigarettes. I constantly had to tell him that he was smoking too much. It wasn't really any of my business except that whenever he smoked, I requested that the window in the truck be opened, but when it was twenty-five degrees outside, the constant cold air blowing in the truck every five minutes became annoying. When I would remind him that he had just put out his last cigarette about ten minutes before that, he would look at me with his head cocked to the side and say, "Really, I didn't realize that." Then he would he tell me how I should be his life coach to help him quit smoking!

The other significant thing about Tim was his desire to become a Jehovah's Witness. He and I would have some fairly heated discussions about religion. After a while, we figured out the common beliefs that we shared and chose to only discuss those parts of the Bible. I don't remember what it was that made him such a unique character—I guess he had a way of telling funny stories as I did—but I remember we had lots of laughs together. The one thing that became evident to me by watching Tim struggle to get recognized for his intelligence, was his main fault was being meek and quiet.

Stoner Kids

We used to get a lot of rodmen/chainmen that would come to work with our crew that were just hired with no experience. I could relate to that practice because that's how I was initially hired, but these kids mostly came from the west end of Hagerstown and didn't know what real work was. These kids also lived in the city and didn't know how to deal with anything related to the outdoors. They were mostly drug-using, lazy kids who spent a year of partying after high school and now were somehow forced to get a job by a stepfather who was tired of their lazy ass lying about the house all day and partying all night. I don't remember any of these kids working their way up into more responsibilities on the crew. They might say that they wanted to learn how to run the instrument, but they really just wanted to make more money and do less labor.

One of this type of guy was named Joe, and he started working on my crew because we had to have a five-man crew for MD SHA jobs. As we drove down Interstate 270 toward Washington, Joe said, "Oh, I was down here the other week on 4/20 with the NORML group." Now, in case you didn't know, NORML stands for the National Organization for the Reform of Marijuana Laws. I asked some questions and he made no qualms about telling me that they sat in front of the Capitol building and smoked weed. When I accused him of being a stoner, he said, "No way." Was I supposed to believe he only did that as a form of protest? I began to find out that he was a stoner without much sense.

When I told him to go down to a traverse point so that I could set a point in between him and the point that we were currently occupying with the instrument. I instructed him to stand on the point and wave his arms when he could see me. I got to a place along the shoulder of an on-ramp to I-270 and looked back at the instrument and I backed away and turned to look for Joe. I couldn't see him, so I continued walking along the shoulder, keeping the instrument in sight. Finally, I saw Joe and he waved at me. I could see the instrument from there, so I set a PK nail in the shoulder and we measured the angle and distance to it.

When the instrument was placed on the newly set point and we tried to turn the angle to the point where Joe had been, we couldn't see it because of the trees in the way. When I asked Joe what he had done, he told me that he moved off of the traverse point so that he could see me! I had to explain over and over how that defeated the whole purpose of the exercise before he could even begin to understand what he did wrong. Then he began to argue that I had not explained myself well enough.

I'm the kind of person who will look for fault in my own actions first and accept whatever blame can be assigned to me, but these kids would make me crazy! I'd begin to question my judgment and even my recollection of what I had told them to do. The other thing these guys would do was forget everything that you'd tell them after the first two items. For example, I'd say, "Let's get shots (points of location) along the center of the driveway, then go over to the fence line and shoot the post along there on one-hundred-foot intervals, then let's get the corners of the house and then those two property corners in the back." They'd get some shots on the driveway and then ask what to do.

Author with Village of Surveyor Sign - 5-30-1988

The Kid from the Wild West

The one memorable kid from the "Wild West End"—as we referred to anything to the west of Burhans Boulevard—was a kid named Brad. This kid had long dirty blonde hair and ripped-up jeans (before that was a popular style). Once he started working on the crew and had access to black magic markers, he would write the names and logos of popular hair bands of the '90s on his blue jeans. I had never heard of Dokken until I saw that written on his jeans. Then he topped if off with some florescent orange paint—just for show. He was just plain stupid! I couldn't even teach him how to perform simple tasks. He just got in the way.

One day as we were picking up the traffic warning signs that notified motorists of a "Survey Crew Ahead," he started holding the stand against his crotch and made some lewd gestures. I told him that I hadn't just witnessed that (in order to claim innocence and ignorance of his actions), but I also told him if he did anything like that again—in the public eye—he would be fired on the spot.

Then one day at the gas station where we had an account and normally filled up, while I was signing the ticket for the bill, he jumped in the work truck and started the engine. This happened to be the former personal Suburban truck of Bill Fox—this had a 454 CI V8 engine with a four-barrel carburetor and a towing package that Bill used to tow a 40' travel trailer. Not only that, but he revved the engine well beyond what was prudent. Also, I should mention this vehicle had Thrush glass packs (loud, high-performance mufflers) on it. It was loud and I just about pulled the kid out and pounded his head. I warned him that he should never do this again.

After leaving the gas station we stopped at the 7-Eleven for coffee and snacks. I almost had to; these guys would be half asleep all morning if they didn't have their coffee. When we went into the 7-Eleven, there was a very attractive woman in there buying something. All of the guys in the store noticed her and made some googly eyes at her.

When I returned to the truck, I noticed that she was in the car next to the work truck. As the guys came out to the truck, each one commented on how pretty this lady was. When Brad came out, he saw us looking at her and decided to make a lewd gesture right in front of her car as she was looking down. She suddenly looked up, saw what he was doing, and then followed him with her eyes as he got in the truck. Brad was laughing it up and looking at us to see if we were all laughing. The other guys were, but I was not- I was fuming.

I had to think quickly because I saw the woman looking at our truck and the phone number was painted right there on the side. I remembered what had happened to the rodman who mooned the crossing guard—see the story under "Part I – Fox and Associates (1979-1981)." I told Brad that he was fired and he challenged me that I didn't have the authority to do that. So, I countered with, "I'll take you back to the office and Tim Witter can fire you."

When I arrived at the office, I told Brad to come in with me to

Tim Witter's office. After explaining the situation, Tim agreed with me and told Brad to be on his way. I had several young guys who worked with me over the years who turned out to be good surveyors, but many did not.

Surveyor in Sneakers

The summer of 1989 was a good year. There was a lot of work and management was hiring new crew members and buying new trucks, only now instead of the old faithful GMC Suburban, we were now getting dark blue Ford passenger vans. This was due to the number of projects that we were doing for the Maryland State Highway Administration with a five-person crew.

Some of the new guys were okay, but some were not. Two of the guys had worked together at their previous employer and they both came over to Fox and Associates. They worked with me for a while so they could get the hang of doing things our way. Jim was already a party chief and his instrument man, John was hoping to be a junior party chief. They were both good ole boys from Frederick County. John had grown up on a dairy farm and gave some indications that his life was pretty hard at home and his dad was the grouchy old farmer that expected everyone to work as hard as him. John was not a hard worker, so I could understand why he was happy to get off of the farm and work on the survey crew.

Jim had his crew and John stayed with me. I was instructed to train him and evaluate him as potential party chief. If he was deemed to be capable, I would get the position that I wanted—to work in the office. I was asked regularly if he was ready to take the crew. I may have rushed him into that position because, honestly, he didn't really impress me with his work ethic.

He could do the job, but he always had a sloppy attitude. He wore sneakers every day and when asked why he said that he couldn't afford good leather work boots. He also had a manner of speaking that just sounded like he was tired. It was actually a slight speech impediment—or perhaps he would have termed it a "spweach impwediment."

He also liked to hang out with some guys who had a rock band of sorts. It seemed he spent a lot of time being a groupie and staying up late with them; therefore, his head and heart were not usually into the work. I eventually answered the inquiry about his ability to be a party chief, with the reply, "I think so, you'll just have to give him a try and see if he can do it."

It wasn't long until I could see that he really wasn't ready, but they needed me in the office and they continued to depend on John to get the work done; but, he didn't. At the end of the day, he would come walking into the office smiling and shaking his head back and forth. If anyone asked him how it was going out in the field, he would continue shaking his head, and then eventually he'd just say, "It was thick, couldn't find any corners," but with his manner of speaking it sounded like "It waath thick, couldn't find any cohnulz." It began to be a habit. Every day he had the same response. I felt like it was a defense or an excuse for his lack of production.

There was one job in particular that this excuse became more suspicious. He kept saying that he couldn't find corners on a job that was an extension of a prior job that had been done several years earlier. Mike somehow got involved—I suppose he was asked about what he had done on the old section when he completed setting lot corners—and confirmed that all of the corners along that back lot line, which was now the common line with the new section being surveyed, and they should still be there.

John was questioned repeatedly about where he was searching without luck, he kept saying there were no corners there. After he left for the job site with his crew one day, Russ asked Mike to go to the job site and show him the corners he had set.

When Russ and Mike got to the job site, the crew wasn't anywhere to be found! They walked back in the woods and found the steel rebars that Mike had set for lot corners years before. They went back to the road and waited. After another fifteen to twenty minutes, John pulled up with the survey truck and crew inside. They jumped out of the truck and began grabbing tools and equipment. Russ asked where they had been and John said they had to leave to go to the bathroom. Russ asked why they had to leave and John told him that he had to "take a sh*t." Russ asked if he had toilet paper in the truck; if he did, he had no reason to leave the job site.

Then he walked him back in the woods and showed him the

corners that were where John had claimed to have been looking. When he couldn't offer a good explanation, he was fired on the spot. They drove him back to the office and told him to get a ride home; his last paycheck would be mailed. That's the way they did things in those days!

The College Boy

One of the most memorable characters that I got to meet while surveying was Craig Angle. Craig was originally from Waynesboro, so we had some things in common to discuss since my mom was from there and both of my parents worked in that town. Craig was not only an interesting person, but he had an interesting background as well. He started by getting a degree in psychology from Western Maryland College (now McDaniel's College) and began working in a local mental rehabilitation institution. He told me later he couldn't stand the electro-shock therapy that they were administering there. He said the doctors would tell the patients that it wouldn't hurt and then he had to assist them in shocking the patients while they writhed in pain.

Since that wasn't his "cup of tea," Craig decided to move to Nashville, Tennessee, and take up photography, and not just casual photography, but taking publicity photos of the country and western music stars! It was my understanding that he had no formal training in photography before this engagement.

He showed me some of the photos that he had taken. They were impressive. He had Kenny Rogers, Dolly Parton, and some others that I don't remember. While he was in Nashville he decided history was interesting to him and he began a quest for truth into the story of the "Great Train Robbery" that had occurred in that region during the Civil War.

Craig began his research and spent years gathering data in order to get the story historically accurate. After his divorce, he moved back home and began taking photos for technical manuals for a local machine manufacturing company in Waynesboro. When he realized that this employment would not allow enough free time to complete his research for what was turning out to be a book, he quit that job and went back to surveying, a job that he had performed during some summer breaks in college.

Craig decided that he'd check with one of his old associates he had worked with during one of his summer assignments—Pat—who was working at Fox & Associates at the time. Craig was hired as a rodman/chainman and began working on one of the crews.

This type of work gave him the flexibility to continue his research for his upcoming book. On days when the rain prevented any fieldwork, Craig would drive to the Bureau of Statistics, in Washington D.C., to research to make sure he had all of the facts. While I had Ed Fleming—described earlier in this section about Fox and Associates—working on my crew as the instrument man, Craig began working with us as the rodman/chainman.

One of the things that irritated me about him was his practice of checking with the instrument man before carrying out my instructions to him. I might say, "Give me fifty feet and then one hundred feet from the gun," meaning measure fifty feet and then one hundred feet from the same point at the instrument. He would usually look to Ed and ask if he should do that or move ahead after the first fifty feet measurement. I would get very mad and remind him that I was the boss and it didn't matter what Ed thought.

Another problem that I had with Craig after I got to know him very well, was whenever we had a chance to talk, we would have some great discussions; and then when I would start to give instructions, on the heels of our discussion, he wouldn't pay attention. He'd end up asking so many questions and I'd have to tell him that I had already told him what we were doing. He responded by telling me it was my fault! He told me I should have stopped and said, "Now we're discussing the job." Other than those two things, Craig was one of the real good guys to work with. He was almost always pleasant to a fault. That is the reason that I assigned to his habit of checking with the instrument man before carrying out my orders. I think he was afraid of offending Ed by not checking with him first.

Another aspect of Craig's personality was diametrically opposed to my personality. He was the most vehement atheist I had ever met. I always thought that an atheist would be a vile, contemptible creature that had hate in his heart for everything and everyone. Craig was a very lovable person that treated everyone with respect and showed signs of joy in his life. I learned a lot from him about how preconceived concepts of people can be wrong.

I challenged him on several occasions about his religious beliefs, but he was steadfast and would not show any signs of waiver or doubt. I once asked him what happens when a person dies. He told me that it was just like a play; the stage lights would go off; the curtain would come down and the audience would get up and leave the theater to go on with their lives. I was shocked and doubtful that anyone could truly believe that way. I countered that he might see things differently when he was on his deathbed. He related a story from his past in which he had been involved in a serious car accident and was rushed to the hospital emergency room. There was a nurse on duty there that knew Craig. When she saw him and looked at his condition, she offered to pass along any messages to his wife or family. Craig said that he was not worried or alarmed; he simply said, "Oh, okay, tell them, 'Bye.'"

I still have a hard time believing every bit of that story because I was raised in a Christian home and was always confident in the truth that when I died I would be taken into a much better place in the presence of Jesus and all of the saints.

While Craig was working on my crew, I started teaching him how to set up a tripod with an optical plummet over a survey control point. This operation involves several steps to have the head of the tripod level, and directly centered over a point. I showed him how to do it an inordinate number of times without having him succeed in doing it correctly. I remember several times after I had demonstrated with careful instruction on how to accomplish the task, he would ask for help and wouldn't even be close to having it level or over the point.

One time I was sure that if I left him to do it, he would work it out on his own. I sent him down through a field and told him to set it up. After about fifteen minutes of fiddling and dancing around with the tripod, he started to holler that he needed help. I walked all the way down to him and showed him how to do it—again.

He kept forgetting one of the steps or getting one of the steps out of order. After many hands-on lessons in *how to set up a tripod* with this intellectual college graduate, who I thought was very intelligent, he asked a seemingly stupid question. He asked if there were any manuals that could be read to understand the process! I laughed at him and told him that there were, but if he couldn't learn from my expert training, then the manual would not help—*besides,* I thought, *it's easier for everyone to learn by doing instead of by reading "how."*

I gave Craig the manual to read and he proved me wrong. After reading through the instrument manual—which I still think is a total bore, giving lots of incidental, useless information—he came to work and set up the tripod flawlessly in about two minutes! By doing that, Craig showed me that not everyone learns the same way. I realized later that he had worked with technical writers in creating operational manuals that went along with his photographs. I found it obtuse, but he insisted that he could learn better by reading than by doing.

After I had left Fox & Associates, Inc., I told Craig that I would like an autographed copy of his book whenever it was finished. His earlier working title was, *To Stop a Heart.* I don't think he ever explained that meaning to my satisfaction and I constantly told him that it was a poor title for a book that was about stealing a train during the Civil War. I think his publisher told him basically the same thing because the final title was, *The Great Locomotive Chase.*

He was faithful in contacting me when it was published, and he delivered a copy to my house (for the paltry sum of $12.95, or something like that). I wasn't home at the time so I didn't get a chance to press him for the autograph. He left it with my wife.

I contacted him later and arranged to have it autographed. I visited him at his parents' home in Waynesboro where he resided, and he willingly signed the book, "To Eric, the comical Party Chief, from the College Boy." That reminded me of the story when he was dubbed, *The College Boy.*

One day during our lunch break when Craig was working with Ed Fleming and me, we were discussing our past. Craig mentioned that he had a degree in psychiatry; to which Ed immediately said, "Oh, a college boy! That explains everything." It was funny because Craig didn't want everyone to think of him as an intellectual and Ed didn't like him because he was smart, but when he found out that Craig had a college degree, Ed could neatly file him under his little mental compartment for "college boys."

While I was visiting Craig to obtain the autograph, I noticed some baseball memorabilia from little league. I asked about it and Craig told me of his successes with baseball at a young age. He said that he had an older brother that could play and was a good hitter. Craig had struggled with connecting the bat to the ball and his brother told him it was a matter of timing—you just had to get your bat into the hitting zone when the ball was coming through there. Craig said that he was a decent musician by that time (around ten-years-old or something), and he realized what role timing had in music, so he tried to apply the same principle.

While he was on-deck or in the dugout he would sing to himself while the pitcher was going through his wind-up and pitch and then he would mark the syllable of the word that was being sung when the ball was crossing the plate! I think the song that he used to demonstrate was a Beatle's song. It was like he would start a familiar measure of the song as the pitcher went into his wind-up: "She was just seventeen ..." and at that point, the swing would have to be started to reach the hitting zone by the time he was singing, "You [know what I mean] ..." He told me he was extremely successful in hitting after that, but he always used that timing mechanism to swing and connect. I doubted it at first, but I believe that he was accurate in his reporting and wouldn't lie, whether it was about the Civil War or his Little League playing days.

Craig left Fox & Associates under some strange circumstances. I was in charge of the surveying in the Frederick office and there was one crew going out in the field from that office. It consisted of Carl, Craig Angle, and another guy who served as a rodman/chainman. This guy was a little on the lazy side; displaying sloth-like characteristics when I observed him. I forget his name. He eventually left and we replaced him temporarily with a general laborer from Manpower.

The first person that was sent to us was a good worker and was easily trained to perform his duties. For some reason, he had to leave and Manpower sent us a new guy. This guy's appearance did not bode well. He came to work on a cool, cloudy day wearing glow lime shorts that were way too short! His whole demeanor seemed strange. I can't describe it past the glow lime shorts except to say that he was scraggly in his grooming and otherwise just seemed a bit odd. I commented on the shorts and suggested that they may not be suitable and perhaps not even in line with our company policy regarding dress code, but nevertheless, he went out to work with the crew. I can't remember if it was the first day that this guy worked with them or if he lasted more than one day, but I think it was that first day that he was sent home by the party chief, Carl.

One quick and funny story about Manpower personnel, which involves my friend, Mike Sanders, and also relates to the "old men" which came from Manpower: Mike's crew, at one time, included an old guy of about sixty-two years who worked on a Maryland State Highway Beltway survey. I asked Mike if he liked the old guy being on the crew. He said, "Yeah, he's alright. We call him "Pappy." Whenever lunchtime is over, I just feel like crawling up on his lap to have him read me a story while I take a nap!" Then Mike gave his staccato chuckle that made it even more hilarious.

Now, back to the story about Craig leaving Fox and Associates because of his party chief. Carl was a bit of a character—just as much as the other guys that have been introduced. He was raised on an orchard that his mother owned along with her second husband. Carl was used to hard work and from that, his work ethic was impeccable. He also saw a lot of laborers come and go at the orchard and felt no remorse over firing someone for not doing their job. This part of his character apparently bothered Craig because Craig was so sympathetic toward people's feelings that he always wanted to give a person a second, third or fourth chance to redeem himself or herself.

Carl had seen enough of this guy and he brought him back to the office to be sent home. I think it was a rainy day and the crew wasn't going to work much anyhow. It was a standard procedure to relieve Manpower personnel from their duties if they were not well suited to the tasks. The managers from Fox & Associates were constantly trying to get the personnel from Manpower that assigned the jobs, to understand what we needed them to do. We would get some sixty-something old men to work with us on a mountain survey, so it was no big deal to call the Manpower office at the end of the day and say, "Send me someone else; preferably, someone who can actually do physical labor and can understand directions." Then the unfit person was told to go into the Manpower office for another assignment—no big thing—we just wanted someone else.

Well, this particular time when Carl came back to say that this new guy wasn't working out, Craig flipped out and said Carl was unfit to supervise and that he was quitting on the spot! I tried to talk to him and soothe his temper, but he was hot. Craig said that if Carl was going to get away with treating that kid so badly, he was leaving; effective immediately. That was the end of Craig's surveying career, or at least the end of his engagement with Fox & Associates, Inc.

Craig is gone now. I recently found his obituary online. It mentioned his work as a publicity photographer in Memphis and his book. It did not mention that he had a successful Little League career as a hitter by his use of the Beatle's songs. It also did not mention his career as a surveyor. He was forty-something when I worked with him and he died at the age of 64. I miss him now because he was truly one-of-a-kind and made me consider a lot of things with his comments and character. He could make me laugh with some of the stupidest comments. One time while we were getting hubs from the back of the truck, he just started singing, "Hub, a hub of burnin' love," as if he were Elvis singing, "Hunk, a hunka burnin' love." I can still see his face and hear his voice as he used to respond to some new information and say, "Oooh, that's really interesting," as he nodded to encourage more information. I really hoped that I could talk to him one more time, about his path to getting his book published as I neared the finish of this book, but I'll never have that chance.

Chance meeting

While surveying in the field, you never know who you might meet. When we worked in a residential neighborhood, there were always some property owners who would come out of their houses to ask about the project. On one such occasion, while working north of Hagerstown in around 1989, a gentleman came over to the instrument and started to ask what we were doing. As I explained it, he nodded and said, "Yeah, that's kinda like when I worked on the underground pentagon." He went on to say that the surveyors were laying out the caverns inside of Raven Rock Mountain to build Site R, or as we called it when I was growing up, "The Tunnel," and they just barely missed meeting the holes being drilled from each side of the mountain.

I told him that I grew up on the south gate entrance to that government installation. He then asked my name and showed some great interest in that, as he nodded and spoke slowly and deliberately, "You're a Gladhill from Harbaugh Valley." I confirmed that fact and he told me that I should talk to his son-in-law, who had bought a Sharps rifle at an auction which had a piece of paper under the butt plate. Written on the paper was a note from James O. Gladhill, telling how he carried the rifle in the "War of the Rebellion" at Sharpsburg and Gettysburg. It also explained that the gun was owned by Andrew Tressler of Harbaugh Valley.

I knew some Tresslers that lived just up the road from my home, so this was interesting news to me. I absolutely wanted to meet this man who owned the rifle. He had written an article in a magazine for collectors of Civil War memorabilia and he wanted to make the connection to some living Gladhills.

This got my wife interested in our genealogy. She traced back some of the family tree from interviews with my parents, grandparents, and other relatives. Some of them didn't like the idea of looking back into who begat who because it turns out that many couples ended up in holy matrimony to satisfy a moral obligation to marry a girl after making her pregnant! Nonetheless, it was good to know that James O. Gladhill was the brother of my great-great-grandfather. With the information that my wife gathered, and with the help of my parents, we organized a Gladhill reunion. It was very popular and some of the distant branches of the family from Frederick County and Montgomery County, Maryland, came and made the connections between our family and theirs. My wife, Trudy would exchange her research with them and we all gained knowledge about the families. All of this was made possible by a chance encounter with a property owner while I was surveying!

It Was Time to Leave

My time with Fox ended soon enough, also. My good friend Mike was my old party chief that had talked me into coming back to work at Fox & Associates. I had gained a lot of experience and while I was working under Mike, I also took the test administered by the Maryland State Licensing Board and became a licensed property line surveyor. Even so, when my assignment changed from managing surveying in the Frederick office to working in the field in the Hagerstown office, Mike was named the chief of surveys and our relationship struggled a bit.

As the Gulf War drew closer, the economy struggled and our work dwindled. More and more, people were laid off and the snow that kept piling up made matters worse. I was a bit worried because I had just settled on our new house in October of 1990, and January of 1991 was looking very glum for my future. After several days of snow and no work, Mike called me at home on a Friday morning and gave me the bad news—I was laid off and there would be no work for me.

That was somewhat understandable, except for the fact that they kept one instrument man and one party chief, both unlicensed and withless seniority than me, to work along with Mike in the field. The other thing that really ticked me off was Russ's refusal to honor the company policy. The policy was to pay the party chief for forty hours even if the weather prevented them from working. Russ wouldn't pay me for the hours that were missed that last week due to snow. He told me on the phone when I called to protest my light paycheck that they couldn't afford to do so. I was so angry, I practically yelled into the phone, "Oh, yes, when the company policy suits you, it must be followed, but if it doesn't suit the company, then there's reason to not follow it!" Again, Russ could not see my point of view.

I was laid off for about six weeks, but I had already sent out some resumés, so the interviews and offers started coming in, but not before I had started to collect unemployment benefits and even food stamps. We had three children now, with the birth of my daughter, Keri. She was born in November of 1988, and we had moved into our new house, so I was somewhat concerned about our financial future until I landed my new job.

Part VI

Buchart-Horn, Inc.
February 1991 – January 1994

I had interviews with Herbert, Rowland & Grubic, and Buchart-Horn within about two weeks of being laid off. Buchart-Horn was the only one to make an offer, and they were a little bit closer to my home (still an hour-long drive). I reported to work in York, Pennsylvania, at 710 Linden Street. That was the headquarters for the General Engineering Division.

I soon learned that Russell Horn started the company shortly after his tour of duty with the US Army in World War II. Russell set up and operated the company like a company of soldiers in the Army; everything was set up into divisions and each division operated out of a separate headquarters, spread out over the city of York. Mr. Horn believed we should be mobile and ready to relocate at any time—just like the army.

My immediate supervisor, Earl Clouser, was a licensed surveyor in three different states. He was very nice and respectful to me, considering me more of an equal since I had achieved my licensed status in Maryland.

My first assignment of the day was to walk three blocks down the street with Earl to 46 South Richland Avenue; the corporate offices of Pace Resources. There I learned that Buchart-Horn was a part of Pace Resources and also filled out all of the necessary paperwork for my employment. Two hours later, I returned to the General Engineering office and met all of the employees, except for the field crews that were out surveying. There were some engineers-in-training, some landscape architects, and one engineer, named Ron, the head of the division. Then my big task was assigned to me: I was given the company manual to read and a book about management, written by the founder, Russell Horn.

Earl was a pleasant guy, easy to work with and everything, but his idea of marketing was to sit in his office and wait for the transportation department to call and request some surveying work! I liked Earl but he was mired in the bureaucracy of the company. He had been a baseball player, so that endeared me to him. Earl had pitched through high school and was in the minor leagues for a short time, so I was enchanted with his stories of traveling on the team bus and some of the players that I barely recognized from their brief stint in the majors.

Earl also taught me a lot about the business of surveying and engineering. He didn't have to do that, but he did because he must have seen some promise in me. I was used to the old way of keeping the lower guys in the dark and not letting them know how much the company was billing the clients. Since we did a lot of work for PennDOT, we had to use their formula for billing rates. There was an allowable overhead multiplier and an allowable profit, and that's how the wages were used to create the billing rate. He showed me how to estimate jobs and write contracts. That experience was valuable as I grew professionally.

The Airport Engineer

Ron was a different type of boss. I never understood what he knew (or more likely—who he knew) that made him a candidate for division chief or department head or whatever they called him. He was a professional engineer as well as a professional land surveyor. I guess he got his surveying license like so many others: by applying as a professional engineer with enough field experience to qualify as a P.L.S. He never treated me with a lot of respect and always tried to be politically correct, which mostly meant he kowtowed to the female landscape architects who were basically *Femi-Nazis*—a Rush Limbaugh term that he coined about that same time in history.

He asked me once if I could plot existing ground profiles from a contour map. I said, "Sure, what scale do you want them?" He seemed puzzled and showed me the mapping. I again asked what scale to use. He instead asked me what I thought it should be! I helped him come to a decision, but then he asked how I intended to do it. I explained in detail, with him interjecting questions like he had never heard of scaling between contours and plotting them on a profile sheet. I said to someone, "Either he's really stupid or he's testing me." They told me he really was that stupid. He regularly displayed his lack of knowledge to me, but people who knew him and tried to be respectful would say that his strengths were his knowledge of airfield engineering.

One of our clients was the Carbon County Airport. We had done numerous small jobs for them with taxiways and drainage issues. Once, Ron went along with the crew and me to this airport and told us there was a problem with an adjoining property owner. We drove to this farm and Ron introduced himself to the owner. The owner led us to the corner of his property where he had recently had a survey performed and iron pins were set on the missing corners.

There was also a plat prepared that showed the airport's security fence was encroaching onto this farm. Ron acted like he didn't know what to do. He stumbled around this pin while he stumbled over his words and then said we would fix it. I didn't think it was a big deal until I realized that Buchart-Horn had prepared a "Fencing Plan."

I never saw such a thing. It was a construction drawing to put out to bid for fencing contractors. The best part of the story is that no boundary survey was performed, yet we made up a plan and actually staked the proposed fence line in the field. This amazed me! The one thing that gives any surveyor concern when performing a boundary survey is what will be built using those marks as a guide. It is typical for someone to want corner pins and points set on line to build a fence line, but in those cases, it's always a little worrisome trying to make sure everything fits so that there will be no encroachment.

In this case, Ron had decided to layout a fence line intended to be on the property boundary line without the benefit of a boundary survey! Then his most outstanding comment to me was, "Why did this surveyor show our fence as an encroachment and list the dimension of the fence over the line?" My thought was, *Where did this guy get his surveying experience?* Then he made the mistake of asking me what I would have done. I, of course, replied that I would have done exactly as the other surveyor had done. I explained the legal and ethical reasons to do that, but I don't think he ever understood.

On that same visit to the Carbon County Airport, Ron wanted us to stake a "Glide Easement" that the airport owned on adjoining property. It gave them the right to cut trees in the glide path of an airplane descending onto the runway. Part of the job was to locate trees within that glide path and get the elevation to the top of the trees.

The path was determined by a slope of a 20:1 ratio from the end of the runway. Ron was trying to figure out what vertical angle should be set on the scope of the instrument which would be set up on the end of the runway. Well, obvious to anyone who has an idea about surveying, you can't set the instrument directly on the ground at the end of the runway, so whatever the height of the scope, it would have to be subtracted out on the tree where the slope path was projected. I suggested we should obtain remote elevations and by measuring the vertical angle to the top of the tree, along with the horizontal angle and distance to the trees that appeared to be encroaching into the slope path. Ron didn't understand how, but he trusted me.

Later when we brought the notes into the office, Ron couldn't understand how to plot these angles. He only had a 180-degree (half circle) protractor, so he thought it wouldn't be possible to plot an angle of 190 degrees! I had to teach him basic drafting skills. I never did figure out why he was my boss. I guess he talked a good game.

Sammy, from Alabammy, Cook me up some Eggs and Hammy!

In the afternoon of my first day at Buchart-Horn, I got to meet another interesting character named Sam Young. Sam came in from the field, stepped into the cubicle where I was hard at work, and while chewing on a snack cake, looked at me and said, "You must be the new office guy." I confirmed his assessment. Then he said, "Oh, boy, more office wussies." So, I said, "Well, if I'm an office wussie, you must be field scum." Sam seemed to like that for some reason.

It wasn't long until I got my chance to work out in the field with Sam and his rodman/chainman, Marshall Woodward. Sam and Marshall continued to call the office people "Office Wussies." I retaliated by continuing to call them "Field Scum." Marshall was a very loud-talking volunteer fireman; in fact, he was the Assistant Fire Chief for one of the fire companies in West Manchester Township at the time. I imagined that his hearing was shot from all of the sirens.

The Author with Sam Young and Marshall Woodward (at Buchart-Horn, Inc.)

Sam and Marshall cursed like a couple of Marine infantrymen. They also smoked a lot of cigarettes and liked to drink beer in the evenings when we worked out of town. This set them apart from me quite a bit as I was trying my best to be a good example for my kids and had abstained from all of that for several years prior.

Sam was quite the jokester. He would engage anyone that he met in a conversation, usually scattered with chuckles and laughter. Most people liked him immediately. The only ones who disliked him were grouchy, hard-core people who didn't even like themselves. His whole appearance was somewhat comical. He was short and stout and had a huge mustache that was auburn. When he wore his normal cold weather outer-clothes—which consisted of a green jacket, green vest, and a green cap—he reminded me of the Nintendo game character, Luigi. When he would dress up and wear a trench coat, it reminded me of a picture that I saw on the front of People Magazine of the comic Sam Kinnison. Maybe I was just struck with the graphic of seeing the name, "Sam" next to a guy with the same build and general facial features as my co-worker, but either way, Sam was a funny character.

Sam's biggest display of his jokes was on April 1, 1993. Since it was April Fool's Day and we were working along the road doing sewer stake-out, Sam was at his finest, trying to pull pranks on everyone that he could. Some examples: When we went into the convenience store for lunch, he went inside and announced that he had just backed into a nice, new red Mustang. When the owner identified herself, Sam simply said, "April Fools." The lady was so relieved that she wasn't even mad. Then Sam stopped someone pulling out of a driveway and said, "Wait, wait—you have a flat tire." Only when the driver stepped out to look did Sam say, "April Fools."

We were all laughing most of the day, and Sam kept getting more and more out of control. Finally, when we were on a back road and there weren't many cars around, Sam got bored and decided to get us rolling again. He picked up a bottle along the road and made some goofy remark about giving someone a flat tire, then he launched it high in the air and it broke right in the middle of the road! We were startled by this act of vandalism and asked Sam what he was doing. He simply said, "April Fools!" I burst into laughter, but Marshall just asked what that had to do with April Fools. Sam said it was a joke on us because it didn't really break. That just made me laugh even more since it was obvious that the glass was scattered on the road and Sam was just being a real goof. I never did know what triggered him that day except that he just wanted to use that day to be a fool.

Sam Young had some interesting sayings that made him stand out in my mind as an interesting character. His expression for a normal general-type of a person whom he didn't particularly care for was "Sandwich-eater." He'd see some driver or pedestrian in public and say, "Check out that sandwich-eater." When I asked what the qualifications were for that name, he'd just laugh and remark how that person just looked like he'd eat a sandwich! He also used to have a phrase from somewhere that he would insert into his conversations like a comma while he was thinking what else to say: "Knowin' that." It was almost as though he would complete his thought which he had just articulated by summing up the situation and confirming that knowledge had been passed to the listener. It came off as a quirky style of conversation, but became humorous to me the more I heard it. For example, he might start telling me about the Sunday evening that had just passed, and he'd say, "I had to work at Wisehaven tending bar, knowin' that, and I didn't get off work until 10:30."

Sam was a Vietnam War veteran who had received a wound while firing mortars during a battle—or fire-fight as I believe they were called. Someone had dropped a round into the muzzle of the mortar gun and when it didn't fire, they dropped another round on top of that which caused the gun to explode, sending shrapnel into the back of Sam's thigh. He could still work and I don't know if he ever got disability pay from the government; I think not. Because of this he was not the most physically active guy and had a slight limp. He had some character flaws too, but he had one of the biggest hearts of anyone I've ever seen. If he liked you, he would do almost anything for you, but first, he had to like you.

One person that he didn't like was Paxton, so there was always some interesting friction going on between them. There's more about Paxton later, but the best story involving their animosity was when Sam had seen too much of Paxton's ignorant actions in the field and he confronted him in the office and told him exactly how he felt. This led to a tearful apology from Paxton and an attempt to hug Sam to make up! Sam would have none of it and fended off the hug with a stiff arm to the chest. This was done in the privacy of the supply room, and I only heard about it later from Sam.

The Fireman

Marshall was quite a character in his own right. He always claimed he was a firefighter by career, but just worked as a surveyor to make money. He could not pass the civil service exam to become a paid firefighter, so he worked on the survey crew and served as a volunteer fire chief for West Manchester Township in York County on his own time.

I don't know why Marshall was so loud, but he was loud at everything. When he talked, his voice boomed. His voice was deep in tone, but somewhat edgy from the cigarettes, so he came off sounding like a talking lounge singer with the volume cranked up to the max. That characteristic made things easier on the party chief whenever Marshall called out numbers to be recorded, but even when he got into the office, you could hear him having a private conversation throughout the office!

It wasn't just the voice that was loud, either. He always had intestinal gas and he released it whenever the pressure built—regardless of where he was. It was common for him to come into the office and let farts that would bubble, flap, and reverberate throughout the office. There were many sophisticated ladies in the office and he never let that bother him a bit. If I gave him a harsh stare or tried to tell him that farting that loud in a professional office wasn't entirely proper, he would just exclaim, "Oh, they don't stink!"

He ignored the fact that it just sounded like someone's gut just erupted and spilled onto the floor! He also had an enormous amount of upper digestive gas in the form of belches, which also erupted at any time that he deemed appropriate—which was really anytime. I saw a good demonstration of how loud he could belch once when we were working on a large farm survey. We had just returned from lunch and I had set up the instrument while Marshall drove around the roads to get to the backsight point across a valley—about one thousand feet away. As I was focusing the lens to see him clearly and getting ready to call on the radio for a sight, I saw him pop his mouth open with a sudden force, as if an invisible flood was rushing from his throat. Only after a few seconds—the time it took the sound waves to travel that distance—did I realize what caused his actions. The sound of a belch was heard rolling across the fields! I couldn't believe his belch was so loud that it could be heard from one thousand feet away!

His sleeping pattern was another thing of wonder to me. Since he was a firefighter, he would respond to fire calls in the middle of the night. Just that fact amazed me because I can barely get enough wits about me in the early hours of the morning to find my way to the bathroom when it becomes necessary. But Marshall would come to work with any amount of sleep. If he needed sleep, he simply caught naps any time during the day that he sat down.

The Institution

Now the guy we all avoided: Paxton. He was not well-liked by anyone. He tried to be a real expert at everything and that earned him the nickname of "The Institute" which was short for "The Institution of Higher Surveying Learning." He reportedly had a bachelor of science degree from Penn State in civil engineering, although no one had seen his diploma, and quite frankly, everyone doubted that he did graduate. He just acted that stupid. He had some knowledge but was never open to new ideas or different ways of doing things.

He had a greater fault than his arrogance—drinking. He was assuredly an alcoholic. He came to work smelling of beer and belching foul smells around the office and especially in the truck. We used to make each other laugh by imitating him as we pretended to belch into our closed mouth and then blow it out over the windshield of the truck. That one used to make the guys nauseous! He had an annoying way of saying, "Yooou don't know!" Marshall liked to mock that style and call him "The Institoooooshun." He would say that in his loudest bellow with the o's emphasized. Sam shortened that name to "The Toot."

Paxton, mostly due to his drinking, would call in sick about three or four days out of every month. Sometimes he really was sick because of his poor health and eating habits. When we would work out of town, we got per-diem money reimbursed to us based on the receipts that we would turn in. Paxton would fake the receipts and buy beer. His food would consist of bologna and cheese and sometimes hard pretzels. He would sit in his room by himself and drink a case of beer in the evening. The rest of us would go out to dinner and have some beer with our meal, but we were eating well every night.

It became a standard menu depending on the night of the week. Mondays were for Chinese food, Tuesdays, pizza, Wednesday was steaks, Thursday was our last night out and it was reserved for seafood.

Poor old Paxton wouldn't even eat breakfast with us. He would order coffee and torture us with his incessant slurping noises. Once when Paxton was working on the northwest extension of the Pennsylvania Turnpike, he developed a case of gout. Earl Clouser visited the site while the crew was working and saw Paxton limping along the shoulder of the road. He was actually pulling the end of the tape by taking a step and dragging his other foot along behind.

Earl must have taken pity on him and saw how he was delaying the rest of the crew and took him home. I felt sorry for him, too. I saw a beer can insulator once with a saying on it that suited him perfectly. I bought it and left it in the crew room for him. It had this printed on it: "I used up all of my sick days. I guess I'll have to call in dead!" Paxton was so mad about it because he knew he was the one targeted. He went around the office and asked who put it there. He had his ideas, but he never knew for sure.

Kenny

One of the funniest and weirdest stories that anyone can ever tell about surveying occurred in my presence while I was working in northeast Pennsylvania on the Lackawanna Valley Industrial Highway. It is now built and known as the Casey Highway (US Route. 6). This was a proposed highway connecting Scranton with the economically depressed regions north of the city. Most of the area that we were surveying was in old strip mines, but there were side roads that connected to the small towns along the route. One of those small towns was named Jermyn.

We had to go in town to survey the connection points and get some detailed elevations where the bridge across the Lackawanna River was planned. I was working with Sam, Paxton, and Marshall. It was supposed to be two crews, but on this day, we felt we could accomplish more by all of us working together. While we had our instrument set up in the town next to a small grocery store, we met a guy named Kenny. This area was the social hot spot in the town, as every person seemed to come and go at that spot and exchange greetings and news (gossip).

Kenny stepped out of a pickup truck in which someone had given him a ride and stood near the entrance to the store and dabbed his chin with a facial tissue to mop up the drool that was emanating from his mouth. He looked a little bit like Andre the Giant—old-time professional wrestler—but on a smaller scale. He was obviously mentally deficient or had some condition that robbed him of motor skills. Nevertheless, he stood there rocking a bit and shouting greetings to anyone who came by. They all cheerfully waved and said, "Hi, Kenny." He was a happy guy and liked to talk to anyone who would stand still.

I could barely understand anything he said, but he was involved in political discussions and any other subject that arose. Because I was standing still near the instrument and taking notes, Kenny came over and pointed at the instrument, looked at me and said, "Crumfoot—beeble capplestand," or something that sounded similar, while he smiled and drooled and looked to me for a response! It was very awkward because he just continued smiling and jabbering while he drooled and shuffled his feet toward us. After a few nods and awkward exchanges of, "Yeah, oh right." I finally could understand some words, so we—Sam joined me by now—started to ask him about news items of the day. It was kind of cruel because we were having fun somewhat at his expense, but he enjoyed the attention.

We found out that he was a liberal and backed Bill Clinton in the race for president. One of the amusing incidents that happened at the store while we were talking was the produce truck that was unloading in front of the store had a bag of potatoes spill out onto the street and break open. It was somewhat amusing to all, but Kenny found that to be funnier than a Richard Pryor performance! He hopped around and picked up some potatoes and put them in his pockets and offered some to us. He tossed some out into the street and yelled to anyone around while pointing at the potatoes! I wanted to calm him and keep him from theft and destruction since it was obvious that he didn't understand what he was doing, but I was laughing too hard. I was afraid that the other people around us would feel like we were encouraging his behavior by laughing but it was too funny; I couldn't quit!

We finished the survey of the intersection by the store and moved our instrument to another control point to continue our topographic survey about a block away. Well, Kenny saw Sam, who was standing near a control point to give a backsight, and walked over to chat with him while Sam waited for us to finish the set-up and call on the radio for a backsight.

Sam called us first and said, "Hurry up, this guy scares me." I told him that Kenny was harmless, but Sam felt uncomfortable with him. When I asked why Sam felt uncomfortable, he told me that Kenny was reaching into the sweat pants that he was wearing and playing with himself! This was unusual and we were not prepared to deal with this type of behavior. We thought it would be best to leave the area and avoid a situation that we didn't know how to handle so we told him we were leaving and started to pack up the truck.

Kenny wanted to know where we were going so we lied and told him that we were going up to the top of the mountain that we pointed to and then we went immediately to the river crossing which was about a quarter of a mile away.

We arrived at the pull-off area near the river and before we could finish getting all of the equipment out of the truck, Kenny showed up and followed us down to the river. We exchanged glances that communicated to each other, "Now what?"

We just started to ignore him as we set up the instrument and began our work. Kenny was content to wander over in the bushes about 20-30 feet away and pull down his pants, presumably to urinate. We had some nervous laughter over this situation, but it soon turned even worse. I glanced over and saw him fondle himself and reported it in a hushed tone to Sam and Paxton who were right near the instrument with me.

Marshall had waded the river in hip boots and was getting some elevations on a gravel bar in the river. He looked across the water at us and then at Kenny. He made a very intent gaze toward Kenny by squatting a bit and craning his neck toward him, then he looked at us excitedly and began making long stroking motions in front of him through the air and pointing intermittently toward Kenny, who was still only ten yards from us. I thought I understood the charade-like hand signals, but I had to look. There was Kenny enjoying himself—masturbating!

Well, this was about all Marshall could stand; he laughed and pointed and was almost beside himself. Paxton glanced and whispered to Sam to look, but Sam just couldn't do it. It was the weirdest mixture of emotions—embarrassment, amusement, hilarity, and fear—not knowing if he would become violent if we laughed too hard at him. As I had said, none of us were prepared to help a person with a disability who didn't really understand what proper behavior in public was.

I presume that he finished quickly because soon he pulled up his pants and walked up the bank away from the river and we never saw him again! For this we were thankful, but we had a story to laugh about the rest of the week; in fact, whenever I talk to Sam, we still have to laugh about that. Our standard line is, "Wow, I know some people really get off on surveying, but not like Kenny did."

The Night Owl

One of the sharp young surveyors with lots of promise was Mark Shutt. The last I heard of him, he was working in the Memphis office for

Buchart-Horn and was managing the surveying operations. I knew that Mark was sharp and he had a combination surveying and forestry degree, but he would come to work looking like he never had enough sleep.

He worked with me a few times, and I could just tell that this guy was tired—I could see it in his actions and hear it when he would speak, which was rare. When we worked out of town, Mark would wake up and ask if we were going to breakfast. When we answered in the affirmative, his reply would be "Just bring me back some orange juice." He was slow to get started.

In the evening, it was a different story. This guy would bring his Sega Genesis video game along and play Street Fighter and NHL Hockey until late in the night. After about 10:00 p.m., he really got going and would talk rapidly and almost incessantly. I got to know him then because he told stories about his brother and friends with much excitement. I came to realize that some people are just wired differently and their body runs on a natural clock that doesn't coincide with the majority of the population. I guess that's why some folks like nightwork.

The CADD Tech Engineer

The one fellow that I really enjoyed working with in the office at Buchart-Horn was Chris Glassmoyer. This was a young, married guy who had graduated from Penn State and still acted like he was a student. He loved to drink and party, especially when the Penn State football team was having a home game. He would tell us the escapades of the past weekend on a Monday morning while he laughed almost uncontrollably throughout the stories. The standard synopsis would go something like, "We booked up there to [wherever], slammed a bunch of beer, booked over to [wherever], slammed some more beers, crashed [wherever], got up and booked over to the stadium, slammed some more beers, booked [wherever] ..." You get the picture. I would always summarize and check his story by repeating, "Okay, so you slammed, booked, slammed, booked, slammed, crashed, booked, slammed, etc., right?" When I put it that way, he always laughed like a wild man. His laughter was one of those styles that made everyone laugh. It always sounded like he was right on the brink of totally losing control and going over the edge into total madness.

Chris and I got along fine because he and I enjoyed sports, mostly the Baltimore Orioles, and modern rock—which included The Cure, Morrisey, and Big Audio Dynamite, to name a few. We worked together on the newest technology at the time: data collector downloading and processing field data to create digital terrain models (DTM). It was a good combination because he had the experience with computer-aided design and drafting (CADD) software and the engineering education, and I had the field experience and knowledge to know how it should work and what the end product should look like.

We worked together with another employee named Bill Chang. Bill was an Asian guy who had an engineering degree from the University of Taiwan and had worked for Bentley Systems, the developer and producer of the software that we were using to interface our data collector files with the Intergraph CADD system.

It was great to work with a guy who had so much knowledge about DTMs, computers, and surveying. He told us stories about surveying highways through the jungles in Taiwan. Even though his accent was very thick, he was intelligent enough to realize that his pronunciation was not correct and that we couldn't understand him, so he would use different words until his meaning became apparent.

Bill was trying to help us get the settings on our software to make sure that the data was converted properly into x, y, and z coordinates. He was asking about the delimiter in the field files, which is a way to separate different lines of data. It can be either comma, space, or tab-delimited. Bill kept saying, "delimiter, delimiter." To us, it sounded like, "Doesn't matter, doesn't matter," because he put the accent on the first syllable and the rest sounded garbled. We eventually got the meaning by him listing the different ways to delimit data. We made fun of him and imitated him behind his back, but we still had respect for his knowledge and appreciated his assistance.

One of the funny conversations involving Bill was his telephone conversation with a technical support person from the Bentley Systems group, which was based in Atlanta, Georgia. We were listening to the tech guy on the speakerphone and he had a strong southern accent, while Bill had this choppy Taiwanese accent. It was an interesting conversation for Chris and me. Bill was inviting him to come to visit and they could tour the Pennsylvania Dutch country. Bill was doing his best to sell the idea and kept talking about the Amish—with a long A. It was hilarious for Chris and me.

Chris and I got set up in this room—actually it was Chris's temporary office while working on a transportation department project; I would just go down there and work with Chris at times—by ourselves with this powerful computer with all the necessary software, a radio and spit bottles. We would listen to either 99.1 WHFS (a modern rock station) or the York College radio station and make requests while chewing Kodiak tobacco and spitting in bottles. It was fun work for a while, but like always; every good thing must come to an end.

Time for Me to Fly

Work started to slow down. I couldn't get a raise to amount to much or get a promotion. Earl Clouser and Ron kept telling me that I would soon be the chief of surveys because they weren't going to work all their lives there. Well, that was true, but one time Earl would say he was only working for five or ten more years and the next time he'd say that he wasn't going to work more than twenty years or maybe fifteen years. I wasn't going to wait around there for fifteen or twenty years just to get the same annual two percent raise that everyone else got—regardless of their performance. At that time, in 1994, we were hosting an exchange student from Brazil, named Renata Crema Velloso Vianna. With her added into our household, we qualified for reduced lunch rates at the school. I guess my annual income at that time was about $23,000.

I also learned there was a former Buchart-Horn employee of Portuguese origin that had started his own engineering company with a Minority Business Enterprise (MBE) designation, which allowed him to obtain contracts with various government agencies (PennDOT, Pennsylvania Turnpike, and FEMA). He needed a surveyor because a lot of his contracts included surveying—something he had advertised having the capabilities to perform, but in fact, did not. That should have told me something about his business ethics, but for some reason, I trusted him and went to work for an extra $0.75 per hour wage.

Part VII

Pinto Engineering, Inc.
January 1994 – March 1995

At first, Eduardo Gomes Pinto seemed like a good businessman and smart boss. He had received his bachelor of science degree in engineering from the University of Paris, where he met his German wife. It reminded me of the lyrics to the Chuck Berry song, "It Wasn't Me," which says, *I met a German girl in England, who was goin' to school in France, said we danced in Mississippi at an Alpha Kappa dance.* He went on to earn his master's degree from Penn State. He interviewed me and told me, in his thick Portuguese accent, "I am not a surveyor, you are a surveyor. I need you to do all of the surveying and run the survey department." I thought it would be a great opportunity to manage the surveying part of all projects done by a company without the headaches of actually running a company.

It didn't take me long to find out that Ed was one of the most ignorant people for whom I had ever worked. He knew how to get work from government agencies, but he didn't know what to do when he got the jobs. He somehow did the engineering on various projects while he was employed at Buchart-Horn, but when he was the owner of his own firm, he expected everyone to do their job with little or no mentoring or input from him. He told me would leave the surveying to me. That wasn't the case at all. As soon as I told him about the additional equipment that was needed, he questioned the need for a shovel to dig for control points. He thought we should be able to use the snow shovels that he had already made a huge investment into, $11 or something!

When we started getting the occasional job that required property line surveys, I told him that we really needed a magnetic locator to find property corners. He couldn't believe that anyone would spend $495 for a magnetic locator to find buried pipes or pins. He couldn't fathom the thought that, without one, a crew would spend hours just to find a pin, when itt would take two minutes to sweep the area with a locator until a signal would indicate a strong magnetic presence. Even worse, we could search for hours, find nothing, and there could still be a corner there unfound, causing flaws in our survey. I could go on and on about Ed, but the only conclusion for the reader would be that he was indeed a very ignorant man.

The Perverted Engineer

Ed had hired this professional engineer by the name of Tim, who was his main project manager. This guy was a little older than me; about thirty-two years old or so, but he had the social graces and overall intellect of a sixteen-year-old boy! He loved to make crude fart noises and unsophisticated, sexual comments. He was well known for standing in an office, having a seemingly intelligent conversation, only to grab his rear-end, make a strangely-contorted face; as if he were suddenly shocked at his situation and then make loud farting noises as he hurried from the room, walking like he had some accident in his pants. Other times he would pick up something off of someone's desk, inspect it, inquire about the ownership and then rub it on his crotch, while he made farting noises! He even stooped so low as to mention my wife's name in his crude comments at times.

Once when I was forced to work out of town for a couple of weeks in a row, I called into the office and he told me that we were, once again, going to be working in Cambria County, staying in a motel room for another week. I told him that I didn't like it and my wife didn't like it either. He told me he would take care of my wife, that she liked when I was out of town and then he could go see her! I hung up on him immediately and when I returned to the office, I wanted to make sure he knew that I didn't like that kind of kidding around. I began to realize I had never clearly voiced my displeasure with his actions. I figured most sensible guys would know that, but obviously, this was no sensible guy!

When I saw him at the office, I told him never to mention my wife again unless he had something respectful to say. He made another vulgar comment, so I told him that if he ever said another crude thing about her, he would be leaving the office someday and an unfortunate accident would occur; somehow his head would be bashed with a wooden stake and whenever he awoke, no one would be around and no witnesses would have seen what happened. He, of course, commented on a veiled threat, so I told him that I was not threatening him, he could just be sure that his head would be split open if he didn't refrain from saying any lewd statements about my wife. He settled down a lot after that, at least with me—but not so with others. He wrote some lewd comments about a secretary and was fired for sexual harassment.

Big Jawn and Shawn

The only thing that made working at Pinto Engineering bearable was the friendships that I had with some of the good people. Some were fun and others were easy to ridicule! In the first group; the fun group, was John Raab, a young engineer in training, recently graduated from a basketball scholarship at Drexel University; then there was Sean Mollohan, a young EIT, as well, a graduate of Northwestern University.

John was a local celebrity in York County Eastern High School, as a basketball star. He was the typical jock, standing six feet and eight inches or so, and quiet, but friendly and humorous. Sean, on the other hand, was just as humorous, but a tad bit more apprehensive in his manner of speech. Sean had served in the Peace Corps in Ecuador immediately after graduating from college. This made him very interesting with an international flair. It made picking on Ed easy because we all had an impeccable imitation of his accent! Sean and John had the whole upstairs of the office, which was actually an old branch bank building - complete with a vault—but that's another story. Ed would call up the stairs to them and no one knew which person he was calling, because his "J" was a soft sound, very similar to the "Sh" sound of Sean.

Ed had two main concerns, neither of which was engineering: grass, especially the dandelions growing in the front yard, and the second was the beverages in the office. He was always looking outside in the spring and summer and asking anyone who would listen whether the grass needed to be watered or mowed, and then he would exclaim,

"Those damned dandelions are laughing at me!"

The other item of concern—the beverages—involved Sean and John. Ed would go to the store and buy sodas, which were provided free of charge to employees, and he'd buy some juice for himself. Then he would tell everyone that the juice was his and not to be enjoyed by anyone else. John would complain, mostly just to ridicule Ed and see what he would do. After a while, Ed would feel guilty and then he would call up the stairs to John in his unique accent, saying, "Jshown, I have something for you." That alone was funny sounding enough to cause giggles throughout the office, but then he would add, "Jshown, I have some jooese." After several pleas from Ed, John would answer, "No thanks, Ed, that's your juice." That just caused Ed to plead more! It was hilarious.

Sean looked like a cross between David Bowie and Hugh Grant, only not quite as handsome as either one. He could always point out something hilarious in an intellectual manner, while all the time, giggling and gasping for breath. Sean would usually keep us interested by telling stories of his life in Ecuador, how they ate guinea pigs, watched American films dubbed with Spanish and with English subtitles, and how they peed in the corner of the pool halls because of a lack of indoor plumbing!

When I started working with Pinto, Sean was the project manager (a term used loosely—he did it because no one else would) on a FEMA flood study and mapping. This was a huge project located north of Pittsburgh, which Ed was proud to have landed, mostly due to his experience and heavily weighed with his DBE status. I think Ed was supposed to be in charge, but whenever Sean asked him for help, he would stutter in his thick accent and say something like, "Ya, ya, ya, J-J-Just geet the sheet done!"

Since this project had a big budget, any overhead time was billed to it. Whenever it snowed, which it did a lot that year, we would all go out to shovel the walk and parking lot. You see, Ed didn't want to pay a snow-plowing contractor $60 to plow it, so three or four of us would take turns shoveling, spending a total of ten man-hours, more or less. Just think, if the average billable rate for those employees was $30/ hour, that only cost $300 to get the snow cleared, but Ed didn't mind; he just billed the project!

Sean would raise his concerns first to his immediate supervisor, Tim, the pervert, but Tim would shrug, make his strangest, goofy expression, and snicker. So, Sean started complaining to Ed about chewing up the budget with these contrived billing hours, saying that every bit of that time was needed to do the engineering, drafting, etc. Ed would usually reply with his normal thick-accented mumbling and slurs which were barely intelligible English, but could be translated into, "Don't worrrrry, guy, you see, eet weell go like queeeck, queeck, a ya, ya aaa ..." Sean would go about doing his best on the project with little or no help from Ed.

We even did fieldwork together, staying out-of-town in the Pittsburgh suburbs; John, Sean, another guy to be introduced later, Kegan, and me. Those are some of my fondest memories, because we really had some fun out of town, going to interesting restaurants where they actually encouraged rude behavior (it said so right on the menu!), going to a Pirates game in the old Three Rivers Stadium, and just drinking beer while watching Beavis and Butthead on TV in the room.

While we were there, the episode of Beavis and Butthead aired where Beavis drank all of Stu's root beer and ate his Ho-Hos and assumed the persona of Cornholio! I'll never forget that, because afterward, when we went to the "rude" restaurant and had a couple of "Bucket of Rocks" (Rolling Rock pony bottles in a bucket), Sean became Cornholio, complete with his shirt pulled up to the top of his head! That was completely and totally outrageous, to see a well-educated, cool-mannered engineer, go into a cartoon character in public!

John got into his share of the entertainment by performing the complete production of the song "Greased Lightning" from the movie, Grease, complete with dance steps and wind-milling arms, in the parking lot. I won't say who drove us back to the motel that night, but I will confess that it was the person who won the chugging contest before we walked out the door, and it wasn't me! God forgive us, for we were fools!

Well, the flood study went on, and eventually, we had something to present to the FEMA officials. Ed encouraged us all to attend and pretend to know something special. He told Jim to talk about layers.

What he meant was a reference to a simple fact of using the correct layer names in the AutoCAD drawings. We all hated to sit in meetings with Ed because he was simply embarrassing. Well, there was a lot more work to be done before they would consider the job acceptable, but imagine this: It was over-budget! Guess who got blamed! That's right, it was Sean. Ed would complain to everyone, in no uncertain terms and repeatedly, that Sean "milked" the job!

We were dismayed, to say the least, especially since Ed had not provided any management, mentoring, or assistance during the project. I confronted Ed with the truth, but he continued to say that Sean had milked the job and he should have had it done. That was the beginning of the end for me because I knew that Sean was one of the most conscientious workers ever and Ed was clearly delusional.

Bradley

Some of the people that worked there that were fun to pick on were Bradley and Dawn. Bradley was an EIT with a degree from Messiah College. He actually had to finish his degree at Temple because Messiah did not have a certified engineering program. Brad was a nerdy character that tried to do the projects that were handed to him, but as with other things there, he didn't get the mentoring or instruction that he needed. Bradley was put in charge of a traffic signal plan with no supervision. It was no surprise when PennDOT rejected the plans repeatedly with more and more comments to address. I didn't have too much to say about his engineering, but he always thought he knew more than the next guy, even though he was inexperienced and ignorant about most things. He had started working out in the field at another firm and therefore had earned himself an opportunity to work in the field at Pinto.

When I started working there, Bradley and I were the survey crew. He was not much fun to work with since he tried to tell me how to do most things. Here I was, a licensed surveyor, and I had this unintelligent, young engineer in training telling me what to do! That was difficult, but then even when we were in the office, he annoyed me. We shared an office and we had a table between our desks. When I would eat my lunch at the table, he would turn his chair around and watch me eat. He would have finished his meal, so his only entertainment was to watch me eat! I hated that and told him so, but do you think he would quit? I could almost stand working with him in the office, but when things got slow for him, Ed sent him out in the field with me. Not only that, but we were working out of town and staying in a motel room with my regular instrument man, Kegan.

The Farm Boy

Now Kegan was another story. He was twenty years old and had just graduated from York Technical Institute with an associate degree in CAD drafting. After hearing about his experience in construction with his dad's home improvement business, he seemed like a natural for surveying. He was big, strong, immature, and hard working. I liked him because he was teachable and was willing to try to learn everything about fieldwork. His knowledge of computers made him valuable, too.

Kegan Brauning with the Chevy Blazer (Pinto Engineering)

The worst thing about Kegan was his driving habits. I think he was just used to thrashing everything that he used. If it had an engine, he ran it at full speed. The survey vehicle we used at first was an old Chevy Blazer that had been Ed's personal truck for years, and then he decided it would make a good survey truck. It was in fairly decent shape and it was four-wheel-drive, so it sufficed. Kegan would drive that thing like he was trying to test its limits! If it was on a highway, he drove like it was the Daytona 500, passing everything and stomping the brakes to get stopped. If we were off-road, it looked like a Chevy truck commercial, where they show it romping up the cinder block mountain in slow motion with the wheels slipping and gripping while Bob Seger's song, "Like a Rock," played in the background.

We worked on several projects in western Pennsylvania, Johnstown area. One time we were heading up Route 219 toward Johnstown right after a snow. We had been in Pittsburgh, spent the night, and then drove back the turnpike and north toward Johnstown. The right lane of the highway was wet and mostly bare, but the left lane had snow and slush covering it. Kegan was driving and we were following a tractor-and-trailer with all of the spray coming up on the windshield.

Kegan got tired of following the eighteen-wheeled truck, so he tried to pass. As the slush slowed us down, he punched the accelerator; the tires spun, and the Blazer fish-tailed. He backed off and pulled back in behind the big truck. I advised him to just stay in the right lane.

After a while, he got restless and started muttering about the spray on the windshield. I told him we were not in a hurry and he'd better just relax and back off of the truck. He sighed extra hard and settled in for about five miles, then decided he had enough and wanted to pass.

As he pulled out, I held my breath. He punched the accelerator and as the Blazer fish-tailed, he steered away from the skid and caused us to spin out! We did two and a half revolutions (900 degrees, if you want to try and add it up) alongside the truck and ended up nudged against the guardrail. He put it in four-wheel-drive, backed out, and got back on the road.

The rush of adrenaline overcame me, and I started cursing him out and telling him to pull over, that I would drive. When I asked what that maneuver was all about, he said that he thought he was supposed to turn the wheel away from the skid to straighten the car out! I was incredulous and got really mad at him for that near-accident. I usually drove after that.

More Bradley

He and Bradley did not get along at all; it was like mixing gasoline and milk! On one memorable occasion, we were working out of town on a bridge replacement project, where the subject bridge spanned a creek that supported beaver colonies. The beaver dams had backed up the stream and created marshland dotted with ponds. It was our job to obtain elevations across this marsh and provide cross-sections and profiles for the flood study and bridge design.

I decided that I would run the 25-foot fiberglass rod with a prism on top in order to get a line of sight over the trees and brush. Kegan was put in charge of clearing line for me—not for a clear line-of-sight, just to get through the brush!—with a machete, which left Brad to operate the instrument. He had done this before and felt confident that he could manage that job.

It was cold and windy with the ponds bearing a skim of ice and a dusting of snow, which blew in my face. We had two-way radios to communicate between Brad, the instrument man, and me, the rodman. I was holding twenty-five feet of telescopic fiberglass rod with a prism on top, waving in the wind and trying to get Brad to follow me to the next shot, focus on the prism and track it with the scope until the infrared signal was returned to the instrument, recording the distance and all necessary data to calculate an elevation. This would normally take three to five seconds. Brad was simply unable to do this. Not only that but he argued that it would not be accurate enough because the rod was waving in the wind too much. I was furious!

I finally told Kegan to go operate the instrument and send Brad down to cut the brush. When Kegan got up to the instrument, he didn't see Bradley. After a while, Kegan saw him walking around a nearby building—he had walked away from the project site! Kegan surmised that Bradley must have had his feelings hurt.

He finally came back to the truck when we were preparing to leave for a lunch break. When I asked him what his problem was, he said there was no problem. He tried to act like nothing was wrong. I continued throughout the day and into the next day to question him and see what I could do to change his attitude, but to no avail.

I returned to the office and told Ed Pinto that I refused to work with him in the field. Ed told me that I should be a team player and use him because he didn't have any work in the office. I replied that Ed should lay him off then. Ed tried to make it go away by telling me to forget about it, and, "I don't want to hear no bad sheet about Bradley!" I put my complaints in writing just so Ed couldn't ignore them anymore. Eventually, Brad was laid off.

The Ladies at Pinto

There were two women in the office at Pinto. One was Dawn, a CADD operator that used to work at Harley-Davidson on the production line but somehow was told that she could earn a higher wage by doing computer-aided drafting. She constantly complained about her low pay, Ed's management skills, and everything else that didn't suit her. She had a speech impediment that caused her to lisp. I don't think we would have picked on her except that she was such a pain to listen to when she was complaining.

She had a boyfriend named Eric, whom she always talked about, referring to him as "A-wick." He was a bartender and that was just the best job in her estimation. She used to brag about how he made a drink called a "Bart Simpson," which she pronounced "Baht Simpson." "It has five different kinds of liquor," she'd say with her big toothy smile.

Kegan used to love to say anything or do anything to make her angry. They had attended York Technical Institute together, and they had a strong dislike for each other. Once, whenever Dawn bought a new Chevrolet and it was parked in the office parking lot, Kegan walked by it and asked, "Do these new Chevys have a gas cap with an inside release?" and pulled the filler cap door open and just left it open. When Dawn saw this, she got really mad and starting cursing and asking "Who the f*** opened my gas cap door?" She was so mad! She said that if she found out who it was she would perform a physically impossible, perverted act! We just had to laugh at her when we heard the profanities that she had let fly.

The other woman in the office was the administrative assistant named Christine. She was a very fun-loving, nice girl who had graduated from Shippensburg University with a degree in communications. She didn't want to stretch herself professionally, so she was content to work in an engineering office and do the clerical work. She did a good job at accounting, payroll, billing, general secretary duties, and receptionist duties, but she hated to make phone calls, even to order pizza for lunch! She would whine and complain and ask someone else to call for the pizza. She explained that she was afraid they would ask her too many questions that she couldn't answer. Here she was a communications graduate and she was afraid to make a phone call!

Ed's Way of Solving a Problem

One of my favorite Ed Pinto stories was about the incident where the water supply line to the downstairs bathroom froze due to very cold temperatures. Ed couldn't understand why the commode would flush but the tank would not refill. I went into the basement with Ed to look where the supply line came into the bathroom. The basement was an unfinished, unheated dirt floor cellar with stone walls. The one-quarter inch copper pipe, which supplied the water, was against the stone wall, right near the top where a cold draft made the air freezing cold. I explained to Ed that we only needed to buy a "heat tape" for about $12 that would plug into an outlet and keep the pipe from freezing for about $0.12 per day.

Ed just couldn't believe that was the best answer. I explained that every mobile homeowner knows that you need to do this to keep the water line thawed in the winter. He protested because it would cost a lot to keep the electricity running on this heat tape! He had a better idea.

The next day, Ed brought in a bottle of windshield washer fluid. He asked if we thought that it would thaw out the water for the commode! We questioned his logic, and he explained that this fluid would not freeze so we could just pour it in the bowl and flush it which would thaw the waterline and keep it from freezing! We just stared and shook our heads. Sean tried to explain that the supply line was frozen and that flushing a fluid that wouldn't freeze just wouldn't help. Ed asked if maybe we could pour it into the supply line then. Sean

Part VIII

Rettew (second time)
March 1995 – March 1999

I had obtained my new job at Rettew without even having an interview. I worked out the details over the phone with Ed Warfel. I think they did ask me to send them a resume, just so they could see what I'd been up to recently. The Friday before I was supposed to start, I went to the Mechanicsburg office to meet the boss of those operations, Gary Stouffer, and his second-in-command, Rick Orner. They told me about the projects which they were working on and told me how glad they were to have me join them. They assured me that I would fit right in, being a hunter, an outdoorsman, and all.

The New Guy works with the Other New Guy

When I arrived to work on Monday, they introduced me to Craig West and said that I would be working with him. He had a tool pouch with a plumb bob, so I assumed he brought it with him. I asked him how long he had been surveying, and he said this was his first day! I later learned that Gary had just given him the tool pouch to use. Oh boy, I knew this would be interesting. I was just a party chief now, not a chief of a whole survey department like before. I was glad to be there because I knew it was a good company and had a lot of interesting projects. Two other licensed surveyors were working in the field as party chiefs; Lynn and Gerry. It seemed like this might be my last stop; perhaps I had "found a home," as we used to say.

Craig and I got along fine. He was a good worker that had already dropped out of Millersville University, worked in a printing plant, and then started working part-time at Print-O-Stat, a store that did reproduction of engineering plans and sold surveying and engineering supplies. He was also taking classes at Harrisburg Area Community College (HACC) for architectural drafting and design. Craig learned that if he worked at Rettew, he could get some drafting experience and they would pay for half of his schooling costs. I immediately told him that he should switch to engineering drafting and design because architecture was lame!

We worked together in the field and Craig began to get more and more experience in the office, working on AutoCAD. Eventually, Craig did switch to engineering design and drafting, and I had the opportunity to teach surveying classes at HACC, with Craig in some of my classes. We had a lot of fun working together in the field. Craig was one of those guys that could make me laugh with a lot of nonsense.

One example of the type of foolishness that I'm referring to occurred one day while driving up Interstate 81 to a job in Fredericksburg, Pennsylvania. We were passing by a construction site where the crews were building a new section of the interstate. There were a group of guys with sections of two-inch pipe, using the pipe to straighten the pieces of rebar that projected from the sides of the concrete paving. Craig kept commenting about what a terrible job that would be, calling them rebar-benders.

Then, when we saw a state trooper sitting along the shoulder of the road, he began to imitate a voice over the radio, reporting a sighting of a rebar-bender at a certain mile-marker. I thought that was funny. Later up the road, I started talking about the guys in school that would smoke a cigarette by sucking in through their mouth while blowing out through their noses. We called that act "hot-boxing." I started to imitate the act by breathing in this manner so much that I started to get light-headed and weave on the highway a little bit.

Craig warned me that I might get pulled over by one of the police and again tried to sound like a cop on the radio saying, "I've got a possible box breather at mile marker 120!" Just the way he said it and the incorrect use of the term totally cracked me up! I laughed harder until I almost blacked out. Craig and I would laugh most of the day but still manage to get a lot of work done. Craig worried that the rest of the survey department would see us laughing together and think we were a bunch of screw-offs. I told him not to worry, we were getting our work done and I had friends in the firm that knew I was a good worker. Later I'll explain a little how Craig's fears were well-founded because the guys working there in Mechanicsburg were a tough lot.

Just to show how paranoid he was, though, I'll have to relate a story that we laugh about to this day. One day we had finished a job and returned to the office. We decided we'd go to lunch together and then return to get another assignment. As we left the parking lot, Craig worried that everyone would think we were getting to be extra-special friends or something. I couldn't understand what the big deal was. I guess Craig had been in different situations at other workplaces that taught him to be careful with friendships and allegiances at work.

Anyhow, as we pulled around the building we saw our boss, Gary Stouffer, driving toward us. For some crazy reason, Craig thought that we shouldn't be seen riding together in the truck to have lunch. He quickly slid down in the seat to hide. Gary stopped to talk to me, so I rolled down my window, as did he. Gary started to tell me about a job that he wanted us to go on and asked where Craig was. Craig slowly slid up into the seat, smiling and waving, saying, "Hey Gary!" He was trying to act like he was picking something up off the floor, but we were both so embarrassed that we always knew Gary was a little suspicious.

Craig and I got the opportunity to work together on a big boundary survey for the National Park Service along the Appalachian Trail in Maryland. The boss sent me because I was licensed as a property line surveyor in Maryland and coincidently, I had previously surveyed some of the adjoining properties for the State of Maryland, Department of General Services (DGS) when I was with Fox & Assoc. and A.E.S.I.

I was excited about going back up into the mountains where I had left my mark ten to fifteen years before and finding some more of the old boundary marks from over a century before. Craig was not so excited as we drove down US Route 15 to spend a week out of town, staying overnight in Harper's Ferry, West Virginia. He asked about the presence of snakes in those mountains. He was concerned about poisonous snakes, especially rattlesnakes. I told him there were some rattlesnakes and the occasional copperhead, but the real concern would be the bees, especially the underground nests of yellow jackets. He assured me that he was not afraid of any bees, but he wondered what we would do if the snakes were so bad that we couldn't work. I assured him that would not be the case.

The first day the weather was rainy, so we did some reconnaissance and courthouse research. The next day was nice so we met with a very pleasant farmer named Richard Pry. He showed us some property corners around his farm and allowed us to drive up into his fields and woods roads. When we got up to the top of the mountain, we finally saw the Appalachian Trail. We were taping distances and finding some metal “T” bars that were set for property corners ten years before. One of them appeared to be missing. We could find the witness blazes on the trees, but nothing was registering with the magnetic locator. Craig insisted that it had to be there.

I went back to where the reel of the tape was lying, about one hundred feet away, while Craig continued to sweep the area with the magnetic locator. Suddenly, I heard him saying, “Ouch, ouch ... ouch!” Louder and louder his cries became! I looked toward him and through the slanting sunlight coming through the humid forest air, I could see bees flying all around him. I shouted, “Run—bees!” He ran toward me—swatting himself and yelling, “OUCH, OUCH!” Once, he fell and hit his knee on a rock and paused to look at it. While he examined the blood coming out of his knee, he slowly and more decidedly said, “Ouch,” then realizing that the bees were still on him, his eyes got wide and he ran faster, yelling, “OUCH, OUCH!”

I caught up to him and started swatting the bees off of his back. I could see dozens. I yelled at him to pull his pants down so he could get the bees out of his pants. When he pulled off his shirt and pants and wiped the last of the bees off, he had more stings than I could count! I had received three stings just by trying to rid him of the stinging pests. Some hikers came by just as Craig was getting re-dressed. We just waved at them and wondered how much of our yelling they had heard. Imagine hiking up the trail and hearing someone yell, “Pull down your pants!”

After I asked Craig how he felt, we walked over to the trail and sat there for a minute to figure out what to do. I asked him about any previous allergies, and he said that a bee hadn’t stung him since he was in elementary school. I tried to convince him that we needed to leave right then, but he said that he just wanted to eat his crackers and drink some water. I warned him that his throat might swell shut or he might lose consciousness, but he said he felt fine. I tried to make him understand that if he went into anaphylactic shock, I wouldn't be able to carry him out of the woods and he might die before I got help to him. He did agree to quit early to get some Benadryl to stop the itching. He slept well that night after swallowing a few of those pills.

Long-haired Pete

While Craig and I worked together, Lynn, in his role as party chief, had a young guy on his crew with a ponytail and a smile like Jerry Seinfeld, named Pete. Lynn treated him like a punk, always making up some tasks to do in the morning, like, to make sure he loaded enough stakes because they (meaning Pete) were going to be pounding a lot of them that day. Pete was good-natured and did as he was told.

Pete was interesting. He would work with me on occasion. When we went to lunch together, Pete was never sure what to eat until I ordered, then he just ordered the same thing. I started to take notice of it, and then I would wait for him to order. He would stand there at the counter smiling and saying, "I just don't know," until I would order, then he'd say, "Yeah that sounds good." or "Okay, I'll have the same!" It didn't take long to realize that he had never been out and about very much.

One day I got a cherry pie, made by Tasty Cake. He asked me if it was good. I told him it was pretty much just like homemade. He said he didn't think he'd like it because he didn't like those maraschino cherries. I tried to explain that it was like a standard, homemade cherry pie; it didn't taste like maraschino. He replied that he never had homemade cherry pie. I remembered some of the other things that he said he never had, and I asked him what kind of food his mother made for him. He told me about his alcoholic father that worked night work and yelled at his kids, and how his mom worked and bought frozen dinners for the kids. I felt sorry for him after that.

The Old-timer Party Chief

The two other party chiefs there were both licensed surveyors and they both were dry, work-oriented, and a little bit conceited. Gerry was an older, more experienced guy of about fifty years of age. He continued to work in the field the majority of the time because he was stubborn and just liked it that way. Gerry was slow and methodical in the field. I didn't really care how fast he worked, but he was always concerned with how many hours everyone was working.

Gerry liked to start at 7:30 a.m., take about twenty minutes to eat an apple and some crackers with peanut butter on them while drinking some water in the middle of the day, but then he'd go right back to plodding along, putting in stakes or shooting some topo.

I worked with him once and the pace was drudgingly slow. He'd get back in the office about 4:45 p.m. and then tell the assistant on the crew to put eight hours on the timesheet. His normal working partner was Tom Prall. Tom was nearly forty and the two of them had worked together for years. They nagged at each other like an old married couple.

Gerry really got me angry one time while I was teaching evening classes at HACC, which meant I had to be back at the office by 4:30 to get some food and drive to the campus to teach. One of the jobs that Craig and I were working on during this period was a cell phone tower above Lewisburg, in the village of White Deer (about 1.5 hours from the office) which took about 5 days to complete.

As I was working on the computer late one day, Gerry asked how many hours I had put in per day on that cell tower job in White Deer. I wanted to ask what business it was of his, but I told him we were working 8-hour days. He started telling me in a condescending tone that if we were driving 1.5 hours to the job site and then driving 1.5 hours back to the office, we were only working for 5 hours each day.

I could do some math in my head, so this was not news to me. I explained about my commitment to teaching, which, by the way, was recommended to me by Gary—he didn't want to do it anymore so he got me to teach his classes. Gerry went on to tell me how he and Tom had worked 12 hour days when they had to drive to the job each day so that they could get a full 8 hours on the job site. I just said, "Well, I'm very proud of you." Then I shut the computer down and left the office.

That's the kind of relationship that I had with Gerry.

The Old-timer Instrument Man

Tom Prall was a treat to work with. He always thought that anyone younger than him, and who was a party chief, must be some kind of college boy that thought he was smarter than Tom. Tom liked to give his suggestions like they were the answer to all of the problems. If the party chief didn't like the suggestion, Tom would arch his eyebrows and maybe roll his eyes a little and say, "Oh well, go ahead, you're the party chief." He may as well have said, "Sure, go ahead, but it's a mistake."

We worked together one day when it was about 90 degrees in the shade by 9:00 a.m., doing construction stakeout at the Hershey High School. Tom was used to being the instrument operator, so I took my usual place as the guy holding the prism and pounding in the hubs.

The first thing we did was stake a curbed island at the entrance. This was right on the stabilized construction entrance, which was built using rip-rap (large limestone rocks) the size of a man's fist. All of the delivery trucks that had delivered materials throughout the construction had driven over this entrance and had pounded the rocks into an eight-inch thick pad of crushed rocks that were harder than concrete. It took me all morning to pound in about twelve to fifteen hubs with guard stakes, because every one of them needed a pilot hole which I created by pounding in a "bull pin," a piece of a jackhammer point.

I took about four or five breaks to get a drink and catch my breath. Each time I asked Tom, "Do you want to set a few of these?" He would just say, "Well, when I work for Gerry, he sets all of the points." I knew better than to ask after a while. By lunchtime, I had blisters all over my hands. I told Tom that he would have to set the hubs in the afternoon. He said, "Well if I get blisters, I guess we'll have to quit." After he got blisters, I knew we couldn't quit, but he wouldn't pound anymore, so I went back to the sledgehammer and kept setting the hubs. Fortunately, the ground wasn't as hard where we were setting points then.

That night, when I thought about what was ahead of me the next day, I cried! I literally had an emotional breakdown. I knew the weather, the job, and the crew were going to be the same and I couldn't help but feel total desperation.

Very early the next morning, I saw that some of the points that needed to be set were already in the ground. Lynn and his crew had set them earlier, but they looked like they may have been disturbed by the equipment. I looked at the one hub and saw the point number was written on the stake. I told Tom to "check number 146." He asked if I wanted him to turn the angle for that point number and shoot the distance. It was so obvious that it would be the only way to check it, that I got really sarcastic in my reply on the radio. I told him, "No, I just want you to come over here and stick your nose on the tack and see if it looks right!" He didn't miss my sarcasm and said, "Let's not get testy."

We did have some good times in the field later, but that day was not a good one. When Tom and I worked together on some of the Appalachian Trail projects in the mountains, he actually claimed to have a lot of respect for my abilities, experience, and work ethic. We were both Orioles baseball fans and loved music, so we did form a good friendship despite our differences.

The Know-it-all Party Chief

As mentioned earlier, the other licensed party chief was named Lynn. Lynn had graduated from Penn State, Mont Alto, back in the '70s and was an avid hiker and backpacker. He was stoic and always acted like he knew more than anyone else. Lynn was quick to point out anyone's mistakes and make elaborate excuses for his own goofs. His other annoying habit was his smarmy tone to his voice. He had a bit of an overbite and that contributed to the raven-like tone that he would emit when pontificating about any topic of which he considered himself an expert—and those were many.

Lynn's usual beginning to a reply—a line which we used to mock him—was, a nasal drawn out, "Nahh ..." while shaking his head in obvious disagreement and condescension. One of the best stories about Lynn also involved the previously described longhair, Pete.

I came into the office one morning and Lynn showed me a plotting of some field location, saying, "Look at that, we have fourteen filler caps out there." I had no idea what he was talking about so I made some inquires about the site and so forth. Then Lynn told me this was an open site, just a field with some fenceline. Pete was running the instrument and data collector and had obviously entered the wrong code into the data collector because there were no filler caps on the site.

Normally when shooting topo, the party chief would "run the

rod," meaning to walk with the prism pole and set it down at the points where a location or elevation was needed. Then the party chief would call out on the two-way radio what the feature was. The person at the instrument would enter a two or three-letter code that would be used to identify the feature. Feature codes would be read by the computer program and would create a symbol or line in AutoCAD to create the field drawing. Some days when the traffic noise was bad, or when there was interference in the radios, it was the duty of the party chief to be sure to communicate clearly what feature was being located. I had my own little fun ways to do that; for example, when I got a shot on a guy wire, I'd say "Guy wire, G-W, George Washington," just so they knew what was being shot.

Lynn's method was to get really close to the microphone and speak loudly, which caused distortion and defeated the purpose. I had experienced that when I worked with Lynn and also heard Kevin Karsnak talk about Lynn yelling something into the radio that sounded like "pizuminous" and then when he looked at the code list, he deduced that it must be "bituminous", which is the correct term for macadam or blacktop.

When Pete came in, Lynn asked him "What happened out there yesterday?" and "You had fourteen filler caps." Pete just shook his head and said he didn't know what had gone wrong. Lynn told him that he should go home and study the feature code list. Pete just said, "Okay."

Pete was working with me that day, so when we left in the truck, he asked me what Lynn was talking about. He said, "Does he call mistakes filler caps?" I explained that Lynn blamed him for entering the wrong codes and that Lynn had suggested that they were actually fence posts that were out in this field. Suddenly Pete's face showed some signs of recognition. He said, "Oh, that's what he was saying, 'FP!'" Pete had not understood Lynn's guttural shouts over the walkie-talkie of FP and thought it was FC—the code for filler cap! Lynn would never consider that the whole thing would have been remedied by telling him that he was getting a location on a fence post and allowing Pete to enter the correct code instead of barking two letters that could easily be misunderstood. Pete didn't need to study the code list, he just needed coherent instructions.

The Foul-mouthed Coal Cracker

Kevin Karsnak was a young guy who was hired at Rettew right after he graduated from Penn State with an associate degree in surveying. He can best be described in simple terms as a foul-mouthed coal cracker, having been born and raised in Scranton. He moved to Mechanicsburg and began working in the field with Lynn after Pete left. Kevin seemed like this situation suited him well because he had worked at a produce storage and delivery company before he came to Rettew, and he could handle all of the physical labor that construction stakeout demanded. He seemed like a nice, hard-working guy, who just had a habit of using the "F" word like a form of punctuation in his sentences.

After a few weeks, I had the opportunity to work with him while Lynn was on vacation. Kevin was supposed to have all of the knowledge of the control points and benchmarks on the site, so I relied heavily on him to show me around. Whenever I asked him where a certain point was located, he would just reply, in his thick coal-cracker accent—which sounded a little like a New Jersey accent mixed with a Pennsylvania Dutch accent—"Over dare," with a gesture of his hand. He was actually signifying that it was "over there," but that's as exact as he would get. I had to question him several times with more detailed inquiries and then I'd get more of the "over dare" replies before I could get an approximate location on the point.

I remember being at the Plainfield Rest Area on the Pennsylvania Turnpike and driving around the parking area trying to determine if the point was inside or outside of the fenced area or behind a barricade before I would stop to get the equipment out of the truck. It became frustrating, but I had to laugh because it sounded like he was trying to be funny. I guess it was just his simplified, truck driver mentality carried over from his days at "The Apple House" (which he pronounced Opple Haus) in Scranton.

Kevin and I got to work together quite a bit on a huge boundary survey, which was performed jointly for the National Park Service and the State of Maryland Department of General Services (MD DGS). This was close to the project which Craig West and I had worked on, but this one was much larger. We needed to stay overnight in a place near Harper's Ferry since it was two hours from the office. It was only a little more than one hour from my house, but I thought it was best to stay down there with Kevin.

NPS Survey on the Appalachian Trail (Rettrew - second time)
Kevin Karsnak (top) Author (below)

It didn't take long for us to meet some interesting characters in the local tavern where we had dinner a few times. There was one old boy named Scotty that took a liking to us and starting chatting with us every time we were there. He was a part-time bartender at the establishment, known as the Mudfort Inn, and it seemed he was there all the time.

Scotty was a friendly sort of guy, who we always looked forward to seeing when we visited the Mudfort Inn. He introduced himself to us one night after one of the locals tried to start an argument with us by questioning the work that we were doing on the Appalachian Trail. The local old-timer asked what we were doing in town, and after I explained our work to him, he started ranting about the government spending all of that money on what he considered useless services. He asked us how much money we were making on this survey. I gave him some answers that didn't satisfy his curiosity. I first said that it was none of his business, so he asked us directly how much we were getting paid and I said, "Enough."

That made him mad, so he asked again how much money I made. I told him, "Somewhere between $5.00 and $50.00 per hour." Then he got really serious and stared me in the eye and asked, "How much money do you make?" I told him it was none of his business and he started to really rant and rave. Scotty stepped in and calmed the old-timer down and asked us to step over to a table where he engaged us in a nice conversation. He said just to steer clear of that old coot, he was just ornery.

We became good friends with Scotty after that. He asked for my business card and I gave him one; thinking it was always good for business to make new acquaintances. He thanked me and said, "Now, I can be a surveyor," and then added, "Pleased to meet you, I'm Eric Gladhill" as he extended the card to an imaginary person who he was greeting for the first time!

He proceeded to produce cards from a deputy sheriff, an insurance salesman, and a realtor. I got pretty scared thinking of how he might impersonate me sometime. Once, Scotty called our office in an inebriated state and pretended to be a property owner, saying that we knocked down his shed while we were doing our survey. He left his rambling (but somewhat funny) message on our general voicemail, and when I was questioned about it, I had some 'splainin' to do!

Once when we were throwing darts while at the Mudfort, we overheard a strange conversation. One of the guys, who was a part-owner of the establishment, was talking with another guy who was saying how he could kill or maim someone with his bare hands. He told how he was scared that someone might tap him on his shoulder and he might spin around and kill him because he wouldn't punch someone, but he might, "Palm a man."

He was talking about hitting someone on the nose with an upward punch with the heel of his hand; which is supposed to drive bone fragments into his brain and kill him. It was weird and it became weirder because he continued saying, "Oh yeah, I won't punch someone, but I might palm a man." After that, when we saw this guy, we called him "Palmaman" behind his back! We never learned his name, but when we saw him, we'd lean closer and say "There's Palmaman."

Tattoo

When the big survey project on the Trail was just getting started, our boss, Gary Stouffer, had hired a surveyor who was licensed in Maryland and had always lived and worked in the Annapolis area. He had moved into York County so Gary hired him and scheduled him to take over this big survey. The Friday before he was supposed to start, this new hire, named Rich Lowe, sent a fax to Gary and explained that he had hurt his back during his last week at his old job and wouldn't be able to work for at least six weeks.

This was supposed to be my big break to move into the office and do more management work and less fieldwork. Oh well, I was going back out into the field full time, and I didn't really mind because this was "real" surveying—boundary locations in the mountains—not the little topo jobs or, even worse, construction stakeout.

After Kevin and I worked on the project for about eight weeks, we were told that Rich Lowe was healed and was coming to work for us and would be taking the lead on this project. Kevin immediately protested and said how we were doing a great job and we didn't need any new guys coming in to mess it up. Not only that but he remembered the injury this guy had sustained and Kevin wondered how he would be able to keep up the pace

I should explain that we got really serious on this project and had purchased Kelty brand frame backpacks to haul our equipment in to the section of the mountain where we were surveying. We had also established GPS control points and had worked out parking arrangements with some of the property owners. Kevin was afraid it would take a long time to break in a new guy.

Well, Rich started and we decided the best way for him to get accustomed was to spend one week in the office getting oriented to the firm and plotting deeds that we needed to continue our reconnaissance.

As soon as we met him, we realized two things that set him apart: he was about 10 years older than my 38 years and he was about six or seven inches shorter than my six-foot frame. Kevin complained all week about this old, short guy with a bad back and how he wasn't "too f***ing happy about dragging this guy around." He was sure that he would be doing all of the heavy lifting and carrying. He also said, "You wait; I'll snap his back the first day he's out here." He also started calling him a nickname that stuck forever. "Did you see that little Tattoo f***er?" Kevin would say—referring to the small person character on the old TV show Fantasy Island, as portrayed by Hervé Villechaize "He's so short, he'll never be able to see out the scope of the instrument!" I had to laugh at his insane ranting.

When Rich, or I should say, Tattoo, came with us to the job site the next week, he talked a lot about the deeds having old descriptions which contained the ancient measurement unit of perch and how they did not mathematically close. I told him that this was perfectly normal for old mountain ground. He also asked what the heck a stone pile was. He never saw one and never heard of one mentioned in a deed for a property corner. I told him that this was also very normal. Kevin and I exchanged glances and I began to doubt his ability to do this job.

At the job site, the first thing I did was walk Rich up in the woods and show him and an excellent example of a stone pile. He just looked at it, shook his head, and asked if that was actually a property corner. I guess a surveyor on the eastern shore of Maryland never got so see one of these!

When we took Rich to the Mudfort Inn with us that first night, we introduced him to Scotty and some of the other regulars. The local old coot who wanted to know my wages and loved to complain about the government—I never did learn his name as I was not anxious to know him—came over to the table where we were all sitting and started a debate with Rich (Tattoo) about the government—either the current president or some recent event. Rich began trying to use logic, which everyone knows, doesn't work with a lunatic.

As it started to get heated, we all slipped away from the table and stood in a corner to watch. When the nutty guy started talking loudly, I suggested that we intervene and stop the madness before it got physical. Scotty said, "No, wait, let's see what happens." We laughed about the possible chaos that might arise from this brouhaha. When the old man with the crazy ideas stood up and challenged Rich, Scotty said, "Okay, let's stop this!" I guess he knew the exact boiling point of the old coot. We jumped in and steered Rich away as Scotty told the old fart to go get another beer and settle down. Rich was a bit flustered and managed to stammer, "What just happened? Is that guy nuts?" Well, of course he was.

Somehow Kevin and Rich managed to work together on the big mountain survey for several weeks together, but I always heard Kevin complaining about the "little dude," or "F***ing Tattoo," when he came into the office at the end of the week. Then I started hearing Rich complain about Kevin to the point that I felt I needed to intervene.

I called them into a meeting room one Monday morning when they were ready to leave to work out of town for the week. We discussed all of their problems. Rich didn't want to give Kevin any credit for what he knew about the job, the instrument, or surveying in general. Kevin didn't think Rich knew much about the type of survey that we were doing, and so they were at loggerheads. I tried to mediate their differences and come to an agreement that they would work together. After a while, they shook hands and started loading the truck with the equipment and supplies needed for the week. Suddenly Rich came into the office to the computer where I was working and laid down the envelope holding the cash for expenses during the week and said, "I don't need this shit!" He turned to leave and I said, "What the …?" with a puzzled look. He said, "I quit." I followed him out of the office to see if he was serious, and I watched as he drove away. So that was the end of Tattoo. Kevin and I finished the surveying job

You're not Passionate Enough

The odd thing about the people that worked in the survey department at Rettew in Mechanicsburg was the fact that everyone felt they needed to work overtime, all the time. Rick, who was the field scheduler, would come in early and almost refuse to leave until everyone else in the survey department had left. This made for a long day because someone was always working late.

One time when I was staying late to get something done, Rick came into the main survey room where I was on the computer and sat there reading a magazine while I was working. When his wife called the office, the lady who was cleaning the office, who was also the receptionist (she did double duty to earn more pay), answered the phone and told her, "Oh, Rick's here, he's just sitting here reading a magazine." Rick acted upset that the receptionist would say that. He took the call and I could tell he was trying to explain himself to his wife by saying that he was trying to do some research and then eventually saying he would be leaving work soon.

One time when Rick was staying late, Gary Stouffer, the director of the surveying department, was there and told Rick that he already had too much overtime for the week. He had to tell him at least three times to, "Go the hell home." Rick finally obliged. I asked Gary what was wrong with Rick that he didn't want to leave work. Gary said, "I think he's afraid to go home!" I surmised that Rick was a true workaholic that didn't feel worthy unless he was at work. Rick and I had our differences, but we eventually became close associates and he helped me to learn more about AutoCAD.

Gary Stouffer encouraged people to work long hours. He felt that anyone who worked less than forty-five hours a week was not "passionate" about his or her job. He told me so. Well, I had to admit that I wasn't passionate about surveying, but I contended that I was a faithful employee who did a good job.

When Craig West and I were working on the Appalachian Trail project, Gary scolded me for not working at least nine or ten hours each day. I explained that we were hiking into the worksite, cutting brush all day and this was July and August with temperatures in the high 80s with high humidity each day. I told Gary we carried as much water as we could along with the essential equipment and supplies and when the water was all gone, we worked another hour or so, but then we started to get exhausted and dehydrated. Not to mention that we couldn't carry much food, so we usually survived on some crackers, jerky, and an apple for lunch. By 3:30 or 4:00, we were starving for fluids and some food.

While I was finishing that first mountain survey in the office with boundary determination, survey report, and plat preparation, I had the opportunity to get two free tickets to an Orioles game. The seats were in the lower box section and it was almost my son Dustin's twelfth birthday. I jumped at the chance even though I knew I would have to leave work a half-hour early. This, of course, was taboo and unheard of! It was expected that you would work at least a half-hour late, but never leave early. I reasoned that I had a few hours of overtime already and I knew that we would meet the deadline for this project, so I asked to leave, explaining the conditions. Gary wasn't there so I made my request to Rick. He grudgingly agreed to allow me to leave.

The next day Gary called me into his office and had a long, torturous talk with me about how I was not dedicated enough. He told me that I was not passionate and I never worked enough overtime. He was worried that I would not finish the job in time, and I realized that Rick was concerned that I would burden him with the last-minute finishing touches to the plats. As I tried to reason with him, he told me that working overtime was not equated to dedication, but he still made it clear that I was not dedicated because I didn't work enough overtime! I told him I had obligations to my family that I would not jeopardize by working extra hours. Gary asked why I hadn't moved beyond the role of party chief by that time, when I was over thirty years old. That really hurt my feelings. I stood quickly and I mumbled something as I left his office. I wanted so bad to quit right then and there, but I had to stay until I figured out what to do next.

Kegan Followed

At one point during this time, I ran into Kegan while shopping in Hanover. He had left Pinto Engineering just before me to work for Penn Township, one of the York County municipalities, which had its own engineering department. Kegan told me he really wasn't making much pay and was interested in changing jobs. He asked if I could get him a job at Rettew. I inquired; Gary Stouffer interviewed him and hired him on the spot. I didn't work much with him, but he did okay there. Most of the guys seemed to like him, except for Kevin. He didn't like being compared with Kegan since Kevin had an associate's degree in surveying and Kegan had a certificate in CAD drafting.

One time when a new instrument was purchased and put in the supply room, Kegan grabbed it and claimed it as being assigned to him. Kevin would have none of it and said that if he got it on any particular day, he'd use it. Well, Kegan was adamant about having it so he marked the instrument box with a "K," using a permanent marker. Thinking back now, that could have stood for Kegan or Kevin, but it made Kevin mad because it was a pretty sloppy letter "K."

Kevin took it with us out of town one time. I say *out of town*, but really, it was only Kevin staying in the hotel; I drove home each night because he was only staying in Frederick, Maryland, almost two hours from the office but only 45 minutes from my house. One morning when I picked Kevin up for work, he was fuming! He sat in his room, awake for most of the night, staring at that sloppy "K" that Kegan had drawn. He finally took a marker and made the crooked letter into an "R" and spelled out Rettew. He told me that will fix that "f***ing farm boy." I couldn't believe I had to work with such sensitive knuckleheads.

That was in the winter of 1998-1999; Kevin and I were working on another mountain survey in Maryland for the state. It wasn't fun, because the cold winds and snow were flying over the South Mountain ridgeline where we were working. I was once again finding the old property corners that I had surveyed, found, or set eighteen years prior. It was amazing to find old wooden traverse stakes that were still there!

One Friday when we returned to the office in January of 1999, the scuttlebutt at the office was that Rodney, a party chief who had been hired specifically to work on highway stakeout, had just quit. He had been working on the construction stakeout of Route 322, the Dauphin Bypass, which was slated to last over two years. Some of the guys told me that I would be the one to replace Rodney, but I just thought that it wouldn't happen since I was busy with the mountain survey.

The next Monday when I arrived at work, several people started telling me that I would be replacing Rodney in about two weeks when he was supposed to be leaving. I still felt like it wasn't real until the boss told me so. When I walked by Gary Stouffer that morning, he just said, "There he is, Mr. Dauphin Bypass." When I turned around and asked what that was about, he just laughed. Then Rick told me that I should get with Rodney when he came in and start looking at the Route 322 job because I was taking over.

I couldn't believe that Gary would just hint at the change in duties and then leave it to Rick to explain. I asked how long I should expect to be there before someone else was found to do that work. Rick said the job was supposed to last two years. I replied that I would only be there for one month and they should find someone else to take over on the stakeout. Rick told me that if they couldn't find anyone, then I would be stuck with it. I repeated that I would be there for four weeks only—because, after that time, I would find another job if they didn't have a replacement for the Dauphin Bypass.

Back to Construction Stakeout

I started the stakeout on a cold, dark January morning, and I hated my life. I realized that I was doing the same type of work that I had done sixteen years earlier and I still hated the circumstances that always surrounded construction stakeout. First, there was always a lot of pressure. Everyone needed their stakeout done immediately, and the surveyor had to be quick and accurate. The equipment and crews were usually sitting on the job site, waiting on the stakes so they could begin their work. The foremen rarely planned ahead, so it was the surveyor's fault if work was paused. The stakes had to be correct, precise, and accurate. If concrete got poured in the wrong place, it was hell to pay to get it removed and replaced. The conditions on a construction site are always the worst. It's either muddy or dusty. Bare earth is not the optimal surface for working.

Secondly, the noise on a construction site is always worse than working on the busiest highway. Finally, it is dangerous! The heavy equipment is always ripping around in a time-is-money pace, so you have to watch out for yourself.

One day as we were setting grade stakes along the toe of a cut slope, dump trucks were lining up to be filled by a huge Caterpillar front-loader. I don't remember the model number, but anyone who has worked on big highway projects will know which one I mean—the one with tires about ten feet in diameter and a catwalk around the enclosed cab. The trucks would come pulling up close to where I was setting stakes and then back into a line to be filled by the bucket, which held about four cubic yards of dirt.

I got used to hearing their engines with the fans whirring close to my back, so I quit turning around to see them. While I was getting a shot from the instrument at one stake, I kept my back to the noise of the engine and fan, simply holding the prism steady on the top of the stake, with the prism pole out to the side. Suddenly, the prism pole was being knocked out of my hands by one of those huge ten-foot diameter wheels! I jumped to the side and looked up at the big Cat loader, which had the engine in the rear. I was thinking the engine that I had heard was the front of a truck and the driver would have seen me clearly, instead the operator of the loader that was being backed up did not see me. When he did see me, he flinched as he realized I was there. I got the chills and then later the shakes when I realized how close I came to being smashed under a large machine. I really hated that job then.

There was a situation on that particular project that caused some trouble for me: There are several mechanically stabilized earthen (MSE) walls on that stretch of US Route 322. We were asked to mark a specific grade on one of them while it was being constructed. I went to the top of the wall and obtained an elevation at that spot by having the instrument operator get a reading and then the data collector computed the elevation and compared it to the design elevation. They wanted to check the design grade for the top of the wall, so I simply painted a "fill" mark (the kind like we would usually mark with a black marker on a grade stake) with spray paint on the outside of the wall to give the construction crew a measurement to the proposed top elevation.

Apparently, that was not the best way to mark that grade for them. They claimed that PennDOT would not accept the wall with paint on the outside. Well, there was stone filled up on the inside, so I couldn't paint it there, and besides, the outside of the wall was facing trees and the Susquehanna River; who was going to see it? The construction foreman said that I may have to clean it off with a steel wire brush. I told him that wouldn't happen and he threatened to tell my boss, Gary Stouffer. I said, "Good, then maybe he'll take me off of this job!"

Gary was so mad about this situation that he asked my old friend, Ed Warfel to come and speak to me about my attitude. Ed was very kind and quoted some scriptures about how we should do all things as unto the Lord. I still complained about this assignment and was told that they needed me to be a trooper and help out the firm by doing this stakeout work. I told him that if I was doing such a favor and making sacrifices, then perhaps it would be nice to be compensated with a pay increase!

When I told Craig about this conversation, he was incredulous that I would ask for a raise in the middle of being chastised. I knew that I was not on good terms with management and it was time to go. Each time that I had to leave one job and search for a new one, I remembered the advice that my brother, Ralph, had given me when I was complaining about a certain situation at my job. He said, "Keep your mouth shut, do your job and find something better. When you have a new job, give your notice and leave without creating a big disturbance."

In the middle of February, I started to file my resume with any company that I thought would give me a chance. I even went looking for work at a firm in Ellicott City, Maryland. That would make my commute about ninety minutes each way. I reasoned that I had done it before, working for Dewey Jordan, and I could do it again. I soon had some interviews and was ecstatic when an old, reputable firm from York, Pennsylvania, that had an office in Gettysburg hired me. Thus, started the next chapter in my career.

Part IX

C.S. Davidson, Inc.
March 1999 – Present

I was interviewed and hired by the Human Resources manager, Linda Davidson, who was the wife of the President and CEO, David M. Davidson, Jr., the grandson of the founder, Carl S. Davidson. I was hired to start a survey department in the Gettysburg office because the engineers there were tired of relying on the York survey crews to get the jobs done. The other person involved in the interview was an engineer by the name of Jonathan Holmes. Jon was the leader of the Gettysburg office, and his input regarding the lack of surveying support from York led to my hiring. The only misconception by Jon was that I would be going out in the field every day and would need a young guy to work with me. My discussions with Linda were that I would be a survey manager and that we would need to hire a survey crew. This was the plan, but I wasn't sure how it would all work out.

At that time there were three licensed professional engineers in the Gettysburg office of C. S. Davidson: Jon Holmes, who did mostly municipal engineering (the main type of client served by the firm), Brian Fincher, who did mostly private land development, and Ken Staver, whose specialty was structural engineering.

The situation there became very interesting within the first six months that I worked there. First, Brian Fincher decided to leave the firm to be closer to his son, whom he shared custody of with his ex-wife. I immediately took over three of his projects that were either land developments or subdivisions with public improvements. I was in way over my head.

Soon Ken Staver was also leaving. By September, Jon Holmes had left and I was asked to be the manager of the branch office. I was flattered and somewhat apprehensive; I had never managed anything larger than a survey crew!

About the same time, or shortly thereafter, we were awarded a contract to design the new Adams Commerce Center, to be located near the intersection of US Route 30 and US Route 15, just outside of Gettysburg. The project team consisted of C. S. Davidson, Inc. and the lead consultant, Herbert, Roland and Grubic, traffic engineer, sub-consultant, and William F. Hill Assoc., sewer and water design sub-consultant. Since I was the lead person in Gettysburg with some knowledge of the project, I was named as the project manager. This was almost overwhelming to me since I had never even been the chief surveyor for such a large project. I remember reading an email from Dave Davidson to Jon Holmes about me, saying that I was "an unknown quantity." I made up my mind that I would make my quantity known very quickly, and I strove to do the best job possible in getting this project organized, completed, approved, and built. It did turn out well, mostly due to the whole team, and I cannot take a lot of credit other than I would like to be remembered as a man who worked hard and gave his best efforts.

Best Boss, Ever

Through the course of the Adams Commerce Center project, I became very well acquainted with Dave Davidson since he was the CEO, and actually overseeing all of my work. He was invaluable as a source of wisdom and expertise, and I came to know him as the best boss ever!

Dave had a calm, kind way of supervising and managing that was more like a good coach or consultant. I learned that he was a baseball fan (some would say "nut"—he's the kind of guy who doesn't feel like he really watched a game unless he scored the whole thing—you can judge) and an outdoor sports enthusiast. Even though he was almost fifty years old, he still enjoyed bicycling and windsurfing.

I learned that he came into the business in a similar fashion as the character of George Bailey in the famous Christmas-story movie, *It's a Wonderful Life.* Dave's grandfather, Carl S. Davidson, who was a surveyor and engineer, started the engineering and surveying practice in Hanover, York County, Pennsylvania, in 1923. His son, David M. Davidson, Sr., followed in his father's footsteps and took over the business upon the retirement of Carl S. Davidson.

David M. Davidson, Jr. was accepted at Lehigh University, but couldn't decide whether he wanted to be a writer or an engineer. He decided that if he were an engineer, he could still write as a recreation, but if he were a writer by profession, he wouldn't have much of a chance to engineer things as a hobby. After receiving his BS degree in engineering, he went for and received his master's degree. A short time after that, Dave and his wife, Linda, were planning to go into the Peace Corps and travel to South America and work in impoverished areas to build better facilities and infrastructure for the people there. While he was away from home, he got the news that his father had died suddenly of a heart attack. He came home and stayed to help out with the family business.

A senior engineer at the firm was named president, but soon Dave was appointed to that position and held it for over twenty-five years until 2008, when the board named him CFO, making way for John Klinedinst to become only the second non-family member to be a president and CEO in the eighty-five-year history of C. S. Davison, Inc.

So, as I worked with Dave Davidson, him being my only boss, I realized one day that even though I had my share of learning contract administration, project management, etc., he never lost his temper or talked in a condescending tone to me. One evening I told my wife, I wasn't sure how to respond to such a boss. I had been working there for over six months and no one was yelling at me! I just wasn't used to having someone treat me with so much respect while putting up with my many shortcomings.

Meeting Greg – Strategic Planning

Just before we had landed the contract for the Adams Commerce Center project, I met our new marketing director, Greg Myers, at a bull roast sponsored by the Adams County Economic Development Corporation (A.C.E.D.C.—the developers of the commerce center) to celebrate their first million-dollar grant for the project. The idea was to let the players in the development know that it was coming to fruition. The president of the A.C.E.D.C., Cathy Cresswell, couldn't tell us that we had the job at that time, but she hinted that it would be awarded to us.

Greg turned out to be a nice sort of guy. I couldn't really get a good read on him, but he told me a little bit about his history. He had worked in various, unrelated jobs before coming to C. S. Davidson; most recently he worked for a trash hauler and knew a lot of the municipal managers.

Later in September 1999, the management announced that we would be having a strategic planning retreat in Lancaster. They sent out a request to select our roommates for the two-night stay. Since I was new and Greg was new, I called him and asked if he would want to room with me. He agreed and we began a good long friendship.

The first night at the Historic Strasburg Inn was fairly uneventful. We had some drinks and dinner and then met our business-planning consultant, Bill Peck. He was a colorful, southern gentleman of advanced years—we guessed he was approaching seventy. Bill had an interesting background; he had been a civil engineer and got promoted into management, eventually a CEO, and was currently on various boards of directors. Some of his quaint, down-home sayings made us laugh and endeared him to us.

After our first meeting with Bill that evening, we retired to the bar where Dave Davidson and I talked about baseball and I got to joke around with Greg Myers and some of the other managers. That evening was fairly mild and we went to bed early. The next day was enlightening, but that evening was a lot of fun. I became good at a Bill Peck impersonation that had everyone rolling. I could get the whole gang laughing by repeating some of his quaint sayings, like, "They ain't but a hunnerd pennies in a dolla." or "Mr. President—Big Daddy—that deeply disturbs me." I became the group clown; a position I usually enjoy.

Greg was originally contacted by one of the engineering managers, Jeff Shue, who knew him from a road running club in York. Jeff and Greg dreamed up this idea on the second night of taking a road trip to some bars that they knew in the area. As soon as we got in Greg's car, he started playing the "Jerky Boys" CD. This was a comedy duo that did recordings of prank calls and eventually some live stand-up and even a movie.

I had first heard this act on a cassette tape in Kegan's car while employed at Pinto Engineering. When I heard them, the one guy was calling a building construction supply store about some broken tiles that were delivered to the job site. He had a strong Bronx accent and said his name was, "Frank Rizzo—open your ears, jackass!" We were all in convulsive laughter and had to stop to relieve ourselves in the bushes several times as a result of beer and laughing so hard! I won't mention the name of the place where we ended our night, but it was in Reading, Pennsylvania.

When we returned to the hotel, the one engineering leader, Danny, invited us to have a drink of Jack Daniels with him in his room. All of us went into his room and had some of his liquor. Now Danny is a real engineer; the kind of guy who always gets straight "A's" in school and is very mild-mannered.

He had started the party before our road trip by pouring drinks in a very gentlemanly manner into the glasses that the hotel provides and asking if each person would like ice and water. After we returned from the bar in Reading, it was a bit more boisterous.

Soon he was insisting that everyone have another drink and then just throwing the cap away and announcing that we were going to drink this whole bottle—I guess it was a liter. It was amazing and hilarious to watch the transformation from a kind, quiet, intellectual person to a loud, obnoxious, and giggling drunk!

He began to challenge people and say, "Come on you pussies, let's f***ing drink!" He soon started asking about some other co-workers who were brothers, saying, "Where are those little pussies? Let's call their room!" He wouldn't quit until someone called and said Danny wanted them to come to his room.

I should explain that these brothers were department managers that worked for C. S. Davidson in separate areas. Paul Sauers directed a municipal engineering department and his brother; Bill Sauers, managed the structural engineering department that included Danny. So, Danny was calling his boss a pussy for not coming down for a drink! I believe that Bill and Paul were asleep and weren't even thinking about coming down for a drink, but while they were on the phone, Danny continued his belligerent ranting and goading them about their not wanting to drink with him. I don't think we finished the bottle but we all went to our respective rooms by about 3:00 a.m.

We were all hurting the next morning, but especially Danny. We had breakfast in the dining room and Danny just got few pieces of fruit and maybe some toast. While most of us were trying to get some nourishment in our stomachs, he just pushed his food with a fork.

Poor Danny had been the volunteer to post flip chart pages on the wall. Bill Peck had asked for a recording secretary to write down our ideas from a "Blue Sky" session. We recruited Linda Davidson to do so. As we gave our ideas Linda wrote them on the big sheets of paper and then she recruited Danny to hang them on the wall. The first day went well because he was not hungover. Well, that morning Danny couldn't keep the paper straight as he tried to tear strips of masking tape to hold them in place. It became comical as we pointed out the slant of the paper sheets. We all knew he was hungover hard, and we couldn't keep from snickering even though our heads and stomachs were hurting badly. He eventually left the room for about an hour and he never did admit to hurling. We had a few other strategic planning retreats after that and each one involved some crazy antics, but none will top that first one.

Strategic Planning II

The second strategic planning event was in Harrisburg and again, we stayed at a hotel and had Bill Peck come and help us to track what we had accomplished and plot where we were going. When Linda handed out the vouchers for breakfast in the hotel, someone yelled, "Don't give one to Danny, he won't use it!" Someone else threw in, "Yeah, he'll just push the fruit around on his plate." This caused some prolonged guffaws by all of us.

We went to dinner at a restaurant/nightclub called Gilligan's. We ended up staying there way too late, and then Greg and Jeff decided before we went to bed that night (early morning) that we would get up early and run to get the alcohol and presumably the hangover out of our systems. I was stupid enough to agree with this idea. I was not a runner. I am still not a serious runner, but I do more now than I did then.

I rolled out of bed when I heard Greg get up (we were roomies again) and started to question my sanity and the true usefulness of this plan. Greg assured me that I would feel much better after I ran a few miles. This was the first mention of a distance. I thought maybe a mile or two would be okay, but I soon realized that these guys were going for closer to three or four miles! They soon figured out that I wasn't a real runner, I had these basketball shoes on an old gray pair of sweatpants like the ones used in the '60s. Jeff commented that his feet hurt just looking at those shoes!

I did okay for the first mile then we started into some hills and I realized that we had to run all of those hills back to the hotel. I tried to think of a "jock" way to say, "I want to walk for a while to catch my breath," but those exact words sounded like a cop-out so I tried to think back to any cool sports-related sayings and remembered how in Little League, if you got hit by the ball in the leg or foot, the coach would just say, "Walk it off," so I said, "I'm gonna walk it off for a while." They looked back over their shoulders and asked, "Huh?" I had to repeat it and I tried to say it more jock-like but I ended up saying, "I'll just walk it off here." They laughed and I just turned around, walked for a while, and then jogged back to the hotel. Later, and even to this day, those guys will remind me of that saying—much to my chagrin.

Strategic Planning III

The third year that we had a strategic planning retreat we went to Fell's Point, Maryland, a little old seaport town, which is really just outside of Baltimore and situated on the harbor. This time we had a different business consultant and some of the guys joked that I should do my Bill Peck imitation and just run the meetings on the first day. We all got a good laugh about that, but since I do enjoy being the jester, I found a mathematical joke that assigned values to a letter of the alphabet and then proved that attitude (when the values of the letters were added) could be the only way to equal 100%.

I put this whole text onto an acetate sheet to be viewed with an overhead projector (this was 2002 and there weren't many PowerPoint slide shows at that time). I somehow arranged to have Dave Davidson allow me to speak even before he introduced our consultant, Dave Wahby. I strode to the projector while pretending to run my thumbs up and down a set of invisible suspenders and I did my best southern drawl in starting the explanation. Instead of one hundred percent, I started talking about one hundred pennies in a dollar and everyone was laughing hard. They barely heard what I was saying because they were just laughing too hard. I glanced at poor Mr. Wahby and he had a very quizzical look—apparently, he didn't enjoy this type of humor and he was expecting to lead this group of apparent goof-offs! We didn't enjoy his style of teaching and leading, and several people claimed that I had ruined his whole presentation by starting the session with that Bill Peck imitation!

Kegan Followed Me - Again

Shortly after I started at C. S. Davidson, I had acquired some projects for the firm, and some of the existing projects needed survey work performed on them. I began to use Carl, my former associate from Fox & Associates, who had started his own surveying firm. It was a small operation, just Carl and a helper, but they were more than happy to do our fieldwork under a sub-consultant agreement. Soon, I heard from my old friend and former associate, Kegan. He was looking for work because he was no longer happy doing the cell tower work at Rettew, which required a lot of out-of-town travel. I recommended him, so Linda Davidson and Jon Holmes interviewed him. He was hired to survey with the existing crews out of the York office, to do inspection work, and occasionally go out in the field with me. He came to work in July and seemed to do a great job.

The guys in York liked him. It soon became apparent that we needed equipment, a truck, and an assistant to work with Kegan. We soon found a young man who had two years of experience working for another firm in York. His name was Ty Martin.

The Gettysburg Survey Crew

Ty was a very quiet guy. He came to work for CSD in November and it wasn't long until we were going full bore in the Adams Commerce Center subdivision. Ty and Kegan worked well together and they seemed to get along, but I couldn't get Ty to say more than two words when I saw him in the office.

One day after Paul Sauers came to the Gettysburg office to work, he was talking to Ty—more like talking about Ty in his presence, rather than talking to him, and Ty was being his quiet self. Eventually, Paul said, "That Ty wouldn't say shit if he had a mouthful!" Well, that made all of us laugh—everybody except Ty—he just glared at Paul.

One other time Paul started making comments about Ty's quiet demeanor; Ty remained quiet until he had enough of it, and very pointedly told Paul to keep quiet or he would come around the table and shut him up! That ended that type of joking.

Kegan and Ty worked together on the Adams Commerce Center. They were doing very well in the field and my only complaint was when I tried to get them to do some office work, Kegan would start a drawing, get annoyed and storm out the door mumbling about having some fieldwork to do and Ty would follow along. By the time I got around to checking on what had been done, I'd see that the drawing was left about half-finished and nobody knew what needed to be done except Kegan.

Kegan was a hard-working man. He grew up working on his uncle's farm and was used to manual labor. I used to brag that if you told him to dig a hole in the paved parking lot, he would ask, "How big?" and then start digging with whatever tools were available. His only shortcoming was that he didn't always pay attention to detail. I brought this up time and time again at his annual evaluation. One year I decided this pay raise should reflect his evaluation and gave him less than usual. I told him we could re-evaluate his performance after another three months. He was not happy.

One of Kegan's main complaints in addition to the pay was the coverage of our health insurance. We were self-insured, which meant we were required to pay for services at the time of delivery and then complete a reimbursement form. It wasn't long until Kegan found another job with better pay and better insurance. He came into my office to tell me about it; it was in Arlington, Virginia! It was an hour and a half drive, at least. His father-in-law had been driving to work at the Pentagon for many years and he thought it was a reasonable move. I asked if there was anything I could do to change his mind. He said, "Not unless you can change the insurance." I wished him well.

I thought it over and the next day I asked if he would reconsider if we offered him more money. I thought I knew the answer, but I wanted to find out for sure. He told me that Paul Sauers had already offered to get him a raise and he had refused it. At this time, Paul had been transferred to the Gettysburg office to manage the engineering department there. I was the manager of the surveying department and this move by Paul made me angry. When I confronted Paul about it, he explained it as his attempt to "help." When I asked him how he would feel if the roles were reversed, he realized how he had overstepped his bounds and apologized. I did appreciate his recognition of an error, so, we all moved past this point and said goodbye to Kegan.

The GPS Expert Who Became a Great Friend

Ty Martin took over as survey party chief and quickly became a dependable and careful surveyor. He started working with rental GPS units. He had very little training but became proficient through research and self-teaching. I kept pushing for our budget to include the cost of GPS equipment, either purchase or lease, but the upper management felt that it was not profitable. But then, suddenly we were given a contract to provide aerial mapping control for a large water line project.

The York Water Company had decided that to get through some of the drought conditions that we had experienced, they needed some other sources of water. They got permission to create an intake from the Susquehanna River along Long Level and pipe it to their reservoir, about seventeen miles away. We were awarded the contract to do the surveying mapping control.

It was quickly decided within our company that it was time to enter a lease-to-own agreement with a GPS vendor. We did the deal and assigned Ty to figure out how to make it work. Ty worked with Craig West on this project. It did not go smoothly. The radio signal from the base to the rover which gave the positional corrections was getting interference from other radio signals in the area, so the effective range of the equipment was compromised.

We had to get the vendor to come to the site and look for possible solutions. He suggested using cell phones to send the correction signal. This was in the early days of cell phones with digital data capabilities. As I worked on trying to get approvals to purchase this new technology, the guys in the field struggled. There were some heated arguments as

Ty had to prove what he had learned about the GPS equipment. In the end, the project was completed. I eventually got the cell phones that we needed by signing a contract without any authorization. That got me a stern reprimand and I had to apologize, but we got the job done.

Ty became the full-time party chief but he had to work with whoever we could get to work in the field. He trained several CAD drafters to perform survey tech (rodman/chainman) work. We also had a few guys who thought they would like to survey and then after weeks of training, they decided they would rather work construction or something else.

During one of the times when we needed another survey tech in the field to work with Ty, I got permission to hire my son Dusty to work on the crew. He had dropped out of college and was without a job. I thought that perhaps giving him some tough manual labor would make him realize that he should be in college, working toward a degree that would give him a better job. Dusty had helped me a time or two on a weekend or doing odd survey jobs around our property, but he was as green as they get when he started working in the field.

I knew at first it would be a tough transition for Ty; he was going to have to train another newbie. Of course, I had my own concerns with having my son working under my supervision. There was the possibility of him failing or causing some serious problems. When I asked each of them separately, I got a fairly good assurance that things were going as well as could be expected. Neither one was ecstatically happy, but then again, it was work.

Dusty told me how Ty could be in a very quiet and foreboding mood and then suddenly become happy, silly, and a lot of fun. I knew they both had a very keen sense of humor and active imagination, so I could only imagine they were going to engage in some of the same fun-filled activities that field surveyors do to make the day less boring, as described in some of my previously mentioned anecdotes.

They had been working on an extensive survey known as an ALTA/ACSM Title Survey. The property was in Harrisburg, which meant they had fifty to sixty minutes of travel each way as they went to the site about six or seven days in a row. We even used a young engineering intern, named Munson, who was working in our office for the summer. As I had discovered in the past, a third guy will always change the dynamics of the crew and especially when they are a bit different; it makes the hijinks become the norm and productivity may even drop at times, but it all works out.

Munson thought he had died and had gone to heaven when I told him that we needed his help in the field! I had to calm him a bit and say, "Look, you'll be cutting poison ivy and briars with a machete." He didn't care, he just wanted to be out in the field. He got a lot of ribbing from the guys because of the character played by Woody Harrelson, in the movie, *Kingpin*, especially when the name was used as an adjective to describe someone who was lost and broken.

Ty told me about a particularly entertaining event while they were going to lunch from that job site one day. Some pretty young girls were making eyes at Dusty while they sat in the car next to the work truck at a red light. Dusty was playing it cool, as usual. Munson was excited and told Dusty that the girls were checking him out. When Dusty looked over with a smile and a nod, the one girl lifted her shirt to expose her breasts. Munson went wild about that.

I always felt like a happy worker is a productive worker. It becomes chemistry like that of a ball team when the guys have a lot of good-natured joking and even spitefulness, but when the pressure is on and the work has to be done, they all pull together to win. This situation was no different and Dusty and Ty were enjoying their time together.

The property was a large apartment complex that was sort of a low-rent area. One day as the guys were sitting in the truck checking some of their work and planning the next step, Dusty spotted a fat cat that they had seen lounging around the property. The cat had crawled into the end of a large storm drainage pipe to seek refuge from the hot summer sun. I'm sure it was comfortable in the shade and being partially underground.

As they observed this and began discussing this lazy, fat cat, Dusty just couldn't stand the idea of a stupid cat being comfortable when he was stuck working in the hot sun. He told Ty, "Watch this." He grabbed some pink spray paint and put on a stalk to get in position above the cat at the end of the pipe. He gestured toward Ty to assure that the cat was still there at the end of the pipe. After he laid down on his belly near the end of the pipe, Dusty made a quick move to swing his arm with the paint as low as he could and sprayed the cat! He and Ty got the biggest laugh from watching that cat run out with a partial pink fur makeover! They saw the cat walking around the complex over the next couple of days and it seemed to parade its color even though it was very wary of them now. I do not condone this type of behavior, and they assured me it was a stray and no injury was caused by this bit of vandalism.

After they finished the title survey, they had a large construction stakeout job to do on one of our developments. Dusty took after his dad on this type of work; he did not like it. I enjoyed pounding hubs as much as I did splitting firewood; it is hard physical labor but it had to be accurate. Dusty was smart enough to take batting gloves along to use when swinging the sledgehammer to protect his hands. He was still very unhappy with this type of labor. I thought, *Now he'll see that he should be in college.* I felt bad that he was miserable, but I also knew that there are surveyors somewhere doing this type of work every day, and also, I had done it and survived.

Ty told me about an episode when Dusty was frustrated with a stake that he was trying to pound into very hard ground. He didn't want to walk back to the truck to get the steel bull point used to drive a pilot hole in hard ground, so he kept pounding on stakes. After he split the third one, he pulled it out and threw it as hard as he could toward the instrument and it landed across the graded road from Ty who was standing at the instrument. Ty checked the distance and it was over two hundred feet! I knew he had a great throwing arm, being a good baseball pitcher, but throwing a 1 inch by 3 inch by 36 inch long oak stake is not easy. Ty was impressed but also amused and a little frustrated by the display of temper and immaturity. I could understand both sides of that incident. I've been that frustrated and also disappointed in others' displays of anger.

Ty continued to be a great party chief, eventually working his way into management, and in essence, replaced me as chief of surveys—now titled Survey Technical Services Manager—as I moved on to Client Manager and later Senior Client Manager. Ty and I continue to be great friends and he considers me his mentor.

Ty and I became great friends as we got to know each other well. His father was a career military man and had a strict manner of raising children and dealing with life. It is so similar to my own father. Ty and I shared a lot of stories of how our fathers were our heroes and how they would prove their love for us in various ways, but they could hurt us with their words and their hands. There was no real physical abuse; it was just a different world then where men used whatever methods they thought were necessary to keep a boy in line and make him into a worthy man.

Our fathers were both the "King of the Castle," and their wives would serve them with meekness. It would make us reflect on certain situations with some shock to realize most men just don't treat their wives that way in this day and age. Ty also had an older brother, so we had that "baby of the family" mentality in common.

The other thing that always attracted me to Ty was his ability to quote the funniest lines from movies, songs, or TV shows, usually, the ones that I had failed to even notice when I saw or heard them. Ty would quote these lines while laughing so hard that sometimes I had to ask what he was even saying! After he repeated them a few times, I would start to catch on to what he was saying, but I still didn't usually know where the line was from.

After I heard him quoting lines from a particular movie—like *Anger Management*, for instance—I would find the movie and watch it. Suddenly the lines from the movie were hilariously funny! Ty was good at making me laugh by getting worked up about something and then blurting out semi-nonsensical stuff due to his state of excitement, frustration, or anger. He would also laugh at me for some of the goofy stuff that I said, so we had this mutual respect for our brand of humor.

My Day in Federal Court

C. S. Davidson, Inc. did all of the survey work for the Gettysburg National Military Park, a division of the National Park Service. It was interesting work and we did such things as an ALTA/ACSM Title Survey of the property where the National Battlefield Tower had stood—south of Gettysburg, between Baltimore Pike and Taneytown Road. This tower had become controversial because many people thought its protrusion on the battlefield was an eyesore. Many others—including me—thought that it offered a unique and special view of the battlefield that could otherwise only be seen from an aircraft. Nevertheless, it was ordered by an act of Congress to be taken down.

Our survey plat became part of the eminent domain court proceedings. We also mapped the areas around the Pennsylvania Memorial, Devil's Den, and the Eisenhower Farm. As a result of our engagement with the National Park Service in this way, and because of my previous survey on the land adjacent to the Appalachian Trail on the lands of Richard Pry and others, while I was employed at Rettew, I was contacted about serving as an expert witness in federal court!

Every surveyor prepares for the day that he will be testifying in court. Sometimes that thought is with much apprehension, as it may be imagined that his day in court will be to defend his work as a result of a lawsuit against him, but sometimes that scenario is imagined with some pride if he is to be called in as an expert witness and does not have to fear for the outcome to have possible financial burdens.

In this particular case, I was being asked to prepare and appear to explain how the boundaries were determined in the previously mentioned survey that I had sealed. It was interesting and exciting for me to get the old AutoCAD drawings and start to prepare a series of exhibits to show the judge how this difficult boundary was determined. I knew that Richard Pry didn't agree on where we placed the boundary, but we had sufficient evidence and records to mark the line partway down the hillside on the east side of South Mountain. Richard had always been told and believed that he owned to the ridgeline. It would have made sense because the ridgeline was the dividing line between Frederick and Washington Counties, in Maryland.

I had several meetings to prepare for the hearing with the directors from the National Park Service and with the attorneys from the Department of Justice. We also met in the field with property owners, and I showed them some of the evidence that was used to determine the location of the boundaries. The federal employees all said that it didn't really matter what the judge determined at the hearing, because they were going to buy the land; this was just to determine who got paid for it. So, even if the judge decided to disagree with my findings, it would not be seen as an error on my part, it would just give them a clear direction, beyond what a survey determined, to purchase and transfer the land to the USA.

As more meetings occurred with property owners, I became aware that an adjoining property owner to the south of Richard Pry had hired their own expert witness, Draper Sutcliff. Mr. Sutcliff was a well-known surveyor in the area and was very involved with and a charter member of the Maryland Society of Surveyors. He had surveyed the property south of the Pry land and had also surveyed a property west of Richard Pry that was bought by Richard's son.

I also learned that they had hired their own attorney, James Demma. Mr. Demma is a well-known surveyor and attorney in Maryland. He had written a book about boundary retracement principles. He also taught seminars regularly; in fact, I had recently taken a seminar that he taught on how to be an expert witness! *This will be fun,* I thought; *I will have my first chance to use this new-found knowledge while being questioned in federal court by the very person who taught me these things!*

One thing that stood out during that seminar was his instructions to only answer the question succinctly when it is posed to you. He even told us that if the question can be answered with a, "Yes" or "No," you should answer with the appropriate word and not add any information to it until they ask you to do so. He even told a humorous story about his one appearance as an expert witness when the attorney who was questioning him asked, "Mr. Demma, what are you?" Even though he knew they wanted him to give his credentials as a licensed surveyor and attorney, he gave the following responses: "Father, husband, lover, reader, cyclist," and then stopped and asked if there were any categories in particular that the attorney wanted to know about!

As I prepared for my day in federal court, I dressed in my best suit and tie, gathered the company's laptop computer, which was loaded with AutoCAD software, and had my drawing file all set up to display the various features which were surveyed on colored layers. I jumped in the 1996 Dodge Stratus that had been assigned to me by the company and headed toward Baltimore.

I went from my home in Fairfield, through Emmitsburg and Taneytown, with no trouble. I was feeling good, knowing I had left the house early enough to arrive well before the appointed hearing time. As I headed down Maryland Route 140 toward Westminster, the car started to make some noise and I noticed some vibration. As the noise got louder, I eased the car to the paved shoulder of the road and realized there was a flat tire on the rear right of the car!

This car had been assigned to me for several months before this trip but I had never thought to look at the spare tire, the location of the jack, or any of the tire-changing tools. Now, here I was in a suit and tie at 6:30 in the morning—still dark as night—with no flashlight and a flat tire. Well, I took off my jacket and managed to get the spare installed on the car without getting my pants dirty. I stopped at the first gas station with a restroom to wash my hands and still arrived at the courthouse on West Lombard Street in Baltimore with twenty minutes to spare.

The proceedings in the courtroom were quite interesting. I had never witnessed such a thing; in fact, the only time I had been in any courtroom prior to that day was at the Adams County Courthouse on a fourth-grade field trip. When it came to my turn to take the witness stand, the Department of Justice attorney asked the judge for permission to have me give my testimony from the defendant's table, where I had connected the laptop to the projector for the audience and a monitor for the judge to view the AutoCAD version of the boundary survey.

The actual survey, which was signed and sealed with my Maryland Property Line Surveyor stamp, was submitted as evidence. The judge seemed to pay attention as I switched views on the drawing to show the various original stones set on the old patent lines from the early 19th Century and how they fit together to position the current property boundaries on the land. I don't really know if he understood what he was seeing, but I had worked out my testimony with the attorneys working for the federal government so they understood exactly how the determinations of ownership were made.

After I had completed showing everything that I needed to show, I sat in the witness box and answered the questions posed to me by the attorneys. When I was being cross-examined by Mr. James Demma, he got a wry smile on his face as he approached me. I had approached him after the seminar and asked some questions, so I couldn't tell if perhaps he researched my name and found that I had taken his course, or perhaps he recognized me, but I could see that he enjoyed this game of question and answers.

As he asked me certain questions about the survey, the monuments that were set and found, I answered as simply and briefly as possible, many times just with a "Yes" or "No." It seemed as though he might be getting a little bit perturbed by the seeming insolence and he started to roll his eyes as he pivoted on his heel and turned away from me and asked that I expound upon my answer and tell the court exactly what the boundary law and case history had to say about natural monuments and their importance in comparison to man-made monuments.

I actually felt a bit of pride and confidence in this setting, and found that with all of his questioning he was pointing out that the natural monument of the ridgeline of South Mountain would hold more weight than the rebars (steel pins) that we set for property corners! I had to admit that the ridge was a natural monument, and then he spun away and as he walked back to the plaintiff table, he said, "No further questions." The Department of Justice attorney wisely asked for further examination and asked some other questions to allow me to explain that our rebar property corners were set based on the natural monuments and the direction and distance contained in the deeds of record for the properties.

Draper Sutcliff took the stand and recounted his survey of the Butler property which adjoined Richard Pry's land to the south. His testimony didn't really refute any of my work. I didn't quite understand what value his testimony held for the plaintiffs. The real turning point of the whole hearing was when Richard Pry and his son, Timothy, took the witness stand.

They told of the history of owning the land. Richard was about sixty-five-years old then, and he testified that when he was five-years-old, he rode a horse to drag logs down from the ridgeline. The somewhat humorous occurrence was when Tim was on the stand and the judge offered anyone else the opportunity to ask questions of the witness. He looked around the courtroom and asked, "Anyone?"

Tim's wife stood and identified herself and said that she would like to ask him some questions. I didn't really think that was too odd; I mean, after all, the judge did invite anyone else to ask questions and he made it obvious that he meant, "anyone," but when Tim's wife made her request, the judge suddenly stiffened and then gave a hearty chuckle. Then the attorneys sort of snickered as if they were privy to an inside joke. I realized the humor when the judge said, "I usually don't allow a wife to question her husband under oath, but if you can promise to only ask questions pertaining to the matter at hand, I'll allow it."

Most of us in the courtroom nervously laughed at that statement as we began to imagine the line of questioning that some women might use on the husband while they're under oath: "Where were you on the night when you claimed to be with your buddies shooting pool!" Mrs. Pry simply wanted Tim to bring up some facts regarding the purchase of the property and the title commitment that they had to guarantee clear title to the land.

After all of the testimony, including my well-rehearsed, technical explanation of my boundary determination, the judge ruled that the two Pry families and the Butlers had rights to the land to the ridge of the mountain by adverse possession. This is a legal term for land ownership that is defined by past court cases and needs to have certain conditions to be valid. Among those conditions are: the claimant has to have "color of title," the possession needs to be open and notorious, the possession needs to be continuous, it needs to be actual and it has to exist for a certain period of time. This really did make sense in light of their testimony. It was also palatable for most parties because most of the land that they were claiming was several parcels that were vacant and title traced back to the heirs of last-known prior owners from the 19th Century.

The other two parcels that extended over the ridgeline from Washington County were owned in title by a farmer who really wasn't sure where his eastern boundary was located and a parcel of land that had been purchased but never surveyed, by the State of Maryland. The DOJ attorney asked the judge for clarification by asking, "Does this ownership by adverse possession include those lands owned by the State of Maryland?" The judge said, "Yes, all of those lands adjoining what is shown as their property on the survey at hand." This was quite unusual for anyone in the courtroom who understands one of the basic rules of adverse possession, that it cannot be claimed against any sovereign body or body politic. This would appear to me to be a landmark case, and I mentioned that fact to James Demma, who at the time had a regularly featured legal article in one of the nationally-distributed surveying magazines. James had offered that he would probably get around to writing an article about it sometime soon. I asked him a few times over the next couple of years, but I never did see the article.

Craig Followed Me

When Craig West came to work at C.S. Davidson, Inc. we needed someone who could do AutoCAD drafting by taking the raw field data and turning it into a final plan. Craig had worked with the software that did a lot of the linework and inserted symbols into the drawing based on the field codes used in the data collector. Craig was interviewed by Dave Davidson and me and was made an offer. When he accepted the offer and he started working with us, Linda Davidson remarked that this was the first employee hired, in a long time, without her consent. She hadn't even met him! At least she trusted my judgment. Craig worked with the field crews to develop a code library to take the field data and produce a drawing that was 70% complete. He also developed some standards that made our field drawings consistent. When we acquired the contract to design the subdivision with public improvements for the Adams Commerce Center, Craig became instrumental in preparing the drawings. More on that project later.

I really appreciated Craig, with his experience and abilities, but sometimes he didn't appreciate me when I made promises for plans to be done, and he didn't have the time to do a complete job on the plans. I was interested in budgets and pleasing the client while he was more interested in striving for perfection. This caused some friction. I tried to alleviate his stress by doing some of the drafting and plotting myself.

One day this friction heated up when I couldn't get the plots to come out with the correct line thickness. When I asked Craig, what was wrong with my settings or methods, he blew up on me and demanded that I stay out of "his" drawings. He said that he was the person who should be working on these drawings and I was just creating more problems by trying to do my own drafting. The good thing about working with guy friends is we usually get over it very quickly. I remember talking with Craig and telling him that he couldn't talk to me that way and I would not allow it. About three hours later, we were talking about work and other things like nothing happened.

As the company moved forward in trying to have consistency in our drawings, we had a lot of CADD committee meetings. Craig always had good ideas and would do a lot of work outside the meetings to make the whole system work. There were other drafters; especially from the structural department—who were prima donnas who didn't do anything to help the process. When Craig saw that they were treated the same as him, a conscientious worker, he was somewhat disgruntled. He eventually got a better offer to return to Rettew and he was gone.

Craig West Working on the Adams Commerce Center C.S. Davidson, Inc.

Killer

After Craig left and we needed another CAD drafter, we had an applicant by the name of Charles Sterner, the same name as the Gettysburg Borough Manager. I soon found out that the borough manager was Charles Sterner III and the applicant was his son, Charles Sterner IV. Before I worked at C. S. Davidson, I had not met a person with Roman numerals behind his name; since working there, I've known three men with the same name as their father and grandfather and three people with the number IV behind their name. I guess Charlie, Sr.—as we began to call him since Charlie I and II had passed on—must have told me about his son and his work experience.

Charlie, Jr. was a bachelor of about thirty-something years old. He had a degree in geology and had worked on the environmental team at Rettew while I had worked there in the late nineties. I had met him once as he was working with my friend and former crew member, Chris Embry. In Charlie's previous employment, he had done some surveying on landfills and then took the data and created 3D surfaces and computed volumes of earth.

He had a lot to learn when he came to work, but he learned quickly and became proficient at drafting and calculations. He was and still is an avid hunter and fisherman. He always told us stories about his hunts and kills.

One of the guys who worked there started to call him, "Killer," because of his affinity to hunt and kill anything in season. Charlie the Fourth continued to work for CSD as a CADD technician and also as a field technician from time to time. I continued to have a good work relationship with his father, Charlie III (or Charlie, Senior), the borough manager for Gettysburg.

Charlie, Sr. began to talk about wanting to leave the borough and wondered out loud if C. S. Davidson would be interested in an old surveyor—he had continued to keep his professional land surveyor license up-to-date. He had started his career as a surveyor, working for the local firm, Gettysburg Engineering Company.

We (mostly Paul Sauers and Dave Davidson) made a deal with Charlie, Sr. to leave the borough and come to work for CSD. As he was finalizing his term at the borough, Charlie began to fall into ill health. Just before he came to work for us, he was diagnosed with pancreatic cancer and only worked for us for about a year. I became good friends with Charlie, Sr. during that brief time. We attended some meetings together and had some good talks as we rode together in the car. It was a shame to see him go so soon after getting back into the private sector and working alongside his son.

The Difficult "Engineer"

There was a certain person who seemed to be my archenemy for a period of time. We'll just call him Mr. Mean. He owned a company in Gettysburg. This company served as the engineer of record for many of the municipalities in Adams County. Even though Mr. Mean was not a licensed engineer, he always had some engineer on retainer in order to keep the "Engineering" in the firm's name and the qualification to be appointed as the engineer of record. He was known for a very hard-nosed approach to engineering reviews as well as percolation tests in his duty as SEO. Many people had some hatred toward him because his reviews either cost them additional money or brought their project to an end because his comments prevented them from getting the necessary approvals.

While we were designing the Adams Commerce Center, we had the plans reviewed by Mr. Mean in his role as the township engineer. He was his usual tough self and brought up comments that were not even based on the township ordinances. He seemed to treat me and my counterpart engineer, Steve Loss, as inexperienced kids who didn't know what we were doing. This, of course, frustrated us and made us angry.

Mr. Mean and the township permits officer met with Steve and me a few times, and they would not only question our design but mock it and make unreasonable demands of us. They queried us on how a certain slope would be mowed because of the steepness. We tried to explain that it was a 4:1 (four feet horizontal for every one foot vertical) slope, which is not a steep slope that would prevent mowing. They still expressed concern.

We were trying to get our plans approved on a fast-track timeline and his demand for frivolous revisions took its toll on our morale. We would make all of the revisions in a short, 2-3-day period and resubmit them, only to have Mr. Mean take his good ole time in reviewing them and sending out new comments. He would sometimes hand me the review letter at the planning commission meeting for me to try and respond on-the-spot. Then he would complain about all of the versions of the plans that he had sitting around!

One day the situation came to a head when we were preparing for another planning commission meeting and still did not have any comments on the day of the meeting. It was a landmark day in the history of Gettysburg. It was 137 years to the day after the final day of the famous US Civil War battle. On this day the Gettysburg National Tower was demolished with explosives.

I had asked for the engineering comments on our plans to be faxed to our office before the planning commission meeting. My family came to my office that day around 4:30 so that we could go to the small ridge near the Cyclorama building to watch the tower fall. At 5:02 pm, cannons fired nearby us, and the tower sunk about fifty feet and then toppled—not what it was supposed to do! They had planned for it all to come down into one large heap.

We went back to the office and I said goodbye to my family as I prepared for the meeting. I checked the fax machine one more time and we still did not have a comment letter. I went to the meeting and found that the township office had received their comment letter by fax with a timestamp in the 4:00 p.m. hour! I stood there, once again, in front of the planning commission, trying to respond to comments as I was reading them for the first time.

When I got back to the office I found the letter on our fax machine with a timestamp in the 8:00 p.m. hour. I was furious. I called Mr. Mean in the morning and asked him how this happened after he had promised to send my letter by fax before the meeting. He got angry and said that I was making false claims against him. I offered that either he was lying, I was lying, or the fax machine had somehow held up the transmission or had the wrong timestamp. He got furious and in a high-pitched whiny tone said, "That hurts," and then hung up! As expected, our relationship was strained, at best, for years after that. We can talk these days without being too terse, but he still tells people about my work with comments like, "I guess it's right, I don't know."

The Adams Commerce Center

As mentioned earlier; at the first reference to Cathy Cresswell, there's more to the story of the design phase of the Adams Commerce Center (ACC). When I had mentioned this project regarding my first meeting with Greg Myers, I was at a bull roast to celebrate the first grant being awarded for construction of the planned commercial development, and the date was somewhere around September 1999. We were awarded the contract as the prime consultant shortly thereafter.

The history of C. S. Davison, Inc. with the design of the ACC went back a few years before when a study by a planning group hired by the Adams County Commissioners determined that a certain farm that was located at the southeast quadrant of the interchange of US Route 30 and US Route 15, just east of Gettysburg, was the prime location for a commerce park. Soon afterward, the county commissioners purchased the property from Frank Ruth and then ordered that a design charrette be conducted.

C. S. Davidson was one of the firms invited to participate. It was at that time Dave Davidson had met the new Executive Director of the Adams County Economic Development Corporation, Cathy Cresswell. When they met again, I was there, at a pre-proposal meeting, along with Jon Holmes from our Gettysburg office. The first thing Dave said after, "Hello," was, "Is this for real this time?" Cathy replied, "Yes, we have some money!" I learned a valuable lesson about land development and dreams in general at that meeting: when money comes into the equation, then things happen. Unless there's money people can have all kinds of dreams and plans but nothing happens until there's money!

Then came the proposal submittal; we used Herbert, Rowland and Grubic (now HRG), to do the traffic studies, signal and highway design, and PennDOT permitting, as well as Wm. F. Hill & Associates to do the sewer and water design. We did the subdivision and design of the streets and stormwater detention.

So, after the bull roast, we were awarded the contract and we began to put together our lot layout based on the previous sketch plan that was developed during the charrette mentioned above. As soon as we got started, we were asked to change our scope of work slightly. A local man who had recently retired as CEO of CSX Transportation, the major railroad on the east coast, based in Jacksonville, Florida, David LeVan, decided to open a Harley Davidson dealership in Gettysburg. He wanted to purchase a two-acre lot at the intersection that formed the northeast corner of the ACC along Route 30.

We were asked to give a separate contract to prepare the subdivision plans to create that first lot. This was a nice opportunity for us to get started right away with a little extra work in addition to the over $100,000 fee we had already landed. As it turned out, I was named the project manager for this job!

As the 21st Century was just beginning, I attended a meeting of the Straban Township Planning Commission, where they were reviewing this subdivision plan to create the lot for the new Battlefield Harley Davidson. The Adams County Office of Planning and Development (ACOPD) was under contract with Straban Township to provide review, comments, and meeting attendance regarding new land development proposals, such as ours. The ACOPD and the Township Engineering firm both did review letters and attended the planning commission meeting. Cathy Cresswell was sharp enough to know that, based on the number of comments on this small, one-lot subdivision plan, it would be best to have their solicitor, Gary Hartman in attendance at the meeting.

When our plan came up on the agenda, the comment letter from the ACOPD was being read by their planner who did the review, Sue. The comments were numerous, as previously stated, but the nature of the comments were editorial and full of opinion. One of their protests was about not having a full sketch plan for the entire ACC submitted to the township prior to this subdivision.

When Cathy attempted to explain that the county government should be well aware of this project at this site because they funded the charrette which had formulated the sketch plan and determined some of the layout of the site. Sue began to argue and really speak to Cathy in a condescending and hostile manner. Gary Hartman suddenly stood to his feet and asked the chairman of the planning commission to take charge of this meeting or he would.

The chairman, Riley, just stammered a bit, so Gary proceeded to walk forward and ask who was this woman harassing his client. Sue stuttered, stammered, and then blurted out a retaliatory remark, "I'm not harassing her." To that Gary said, "Yes, you are; I just sat here and watched it. Now who are you?" Sue could scarcely contain her emotions as she blubbered her name and title. Gary again turned to Riley and said, "Mr. Chairman, you need to take charge of this meeting and ask this woman to sit down!"

As Riley asked her if we could dispense with the reading of the letter, she said "We stand by our comments" and just slunk away. I was standing before the commissioner's table at the front of the room feeling a bit excited, anxious, nervous, and a bit scared of what would happen next. Riley turned to me and said, "Well, I guess you can continue addressing these comments." As I tried to speak with a quavering voice and holding the comment letter with shaky hands, I managed to refute some of their ridiculous claims. The plan went on to be approved and the work on the rest of the ACC began in earnest.

Pella Windows Came to Town

One afternoon as I was working on the preliminary design of the lots and roadway for the commerce center, I got a call from Cathy Cresswell. She asked what I was doing that afternoon. I thought it was a check-in to see if I was working on her project, but when I answered her, she quickly got to the reason for her call. There was a carload of men heading toward Gettysburg who were looking for a site on which to build a new manufacturing plant. All that they would say is that they were coming from the west and this was top secret. They wouldn't give their full name, but they wanted to visit the property where the ACC would be built, and they wanted to meet us there that afternoon.

Cathy picked me up at my office at about 3:00 p.m., and we met them near the site and drove back on the farm lanes to have a look at the property while the sun was setting on the horizon. The men offered that they would probably be in touch.

Several weeks later, Cathy sent a message to me and some of the board members of ACEDC and told us that the same group of guys—plus some others from their company—would be visiting Gettysburg and wanted to talk to us about their possible project. They referred to their group as PEPSI, but they assured us it had nothing to do with the soft drink nor any type of beverage. The only thing that they would divulge is that they may be using wood as a raw material in the manufacture of their product.

Eventually, as we had more meetings with the various stakeholders in the community, we had to sign a confidentiality agreement and they revealed that Pella Window was contemplating their first factory east of the Mississippi River and Gettysburg was in the running! The acronym PEPSI was the Pella Eastern Plant Site Investigation. All of us who were involved were very excited. There were elected officials, real-estate developers, utility representatives, and various board members of ACEDC and then me, the professional land surveyor, representing the civil engineering firm contracted to design the commerce center, who only a year prior was doing construction stakeout!

Of course, I quickly got into the business development mode and began to inquire as to who would be doing the site design for the new factory. I had the opportunity to work with one of the engineers from Pella in our office while still keeping anonymity for him and the company. This was the year 2000 and we didn't have Wi-Fi in our office. When the Pella engineer wanted to connect to the internet he had to do it in a public area and had to cover the Pella logo on his retractable internet cable with tape so that none of the C. S. Davidson employees would know who his employer was! These were some exciting times.

I was working ten to twelve-hour days and really enjoying it. My wife, started to voice her concerns that I was becoming a workaholic. I promised her that I was doing this for one project only because it was necessary and that after these six months, I wouldn't work those kinds of hours again. There were some nights when I couldn't sleep because I was thinking about the project, so I got up one morning at about 3:30, showered, and went into the office to work out the design on the roadway to avoid as much of the wetlands as possible. I still think of that when I drive along ProLine Place where the road has a seemingly unnecessary curve to the left and then back to the right to put the road back to the initial direction. I may be the only one who thinks about the large piece of wetlands avoided by making that slight chicane.

Even though the work schedule was grueling and the deadlines were nerve-racking, we managed to have some fun. Craig West was doing the AutoCAD drafting and some of the design along with Steve Loss, a professional engineer with experience in stormwater management. The three of us worked well together and became a close-knit team. One Friday afternoon after everyone else in the office had left for the weekend, Steve Loss and I were still frantically working, pacing quickly from our respective desks to the printer, copier, and/or large format plotter to pick up documents and then pace back to our desk to put them in a folder or an appropriate file.

As I passed Steve in the reception area and realized it was after 5:00 p.m., Steve looked at me and said, "What are we still doing here, Glad (his favorite nickname for me), everyone else is home or at the bar and we're here working like what we do right now is so important? Why don't we just do this on Monday morning?" I pondered that and said that I was just trying to do my best to stay ahead, but it probably wasn't good enough.

The next time we passed, Steve said, "Yeah, we gave our all, but

I guess it's not good enough." This reminded me of the James Ingram song, "Just Once." I started singing in my best voice, which was still probably way off-key, "I did my best, but I guess my best wasn't good enough!" Steve laughed hard and tried another verse; "I gave my all but I guess my all wasn't good enough." We cracked each other up by trying to remember the words to this sappy song which neither of us liked. Since the office was otherwise empty, we just sang loudly enough for the other one to hear from his corner of the building. We sang horribly and as sappy as we could while we took liberties with the original lyrics; "Just once, can't we find a way to get our work all done!" After that, when we were working at a frantic pace, I could always make Steve laugh by singing "I gave my all ..."

Workin' Hard and Playin' Hard

As we eventually acquired the contract to do the site design for Pella, along with the subdivision with public improvements of the commerce park for ACEDC, we had the opportunity to work with the architecture firm of Savage-VerPloeg, based in Des Moines, Iowa. They had designed all of the modern Pella plants and they were going to do this one as well.

The lead architect was a much laid-back man by the name of Steve Gray. His manner and way of speaking was slow. When I would speak to him on the phone, I was always impatient as I waited for him to finish his thoughts or ideas for the design of the site where it related to the building, such as the main entrance and drop-off area in the front. I expressed these feelings to my project team, and when we had him on speakerphone, with Craig and Steve Loss present, they would do things silently in an effort to make me laugh.

I would try to focus on the slow delivery of the architect as he would sometimes change his mind and try to discuss his latest sketch he had sent by fax. After a while, they would go to more extreme measures to get me to crack. I was very amused but was doing my best on one of these calls to remain stoic and listen to Mr. Gray. Just to be extreme, Craig did a headstand on a chair in the office where we were having the conference call! That was it—I broke—right in the middle of a long, slow sentence being delivered over the phone from

Des Moines, I laughed loud and hard! Steve Gray paused, listened for a second and then went on to explain the radius of the curve and diameter he thought would be aesthetically pleasing while I covered my mouth with one hand and motioned dramatically with the other one for the two of them to leave. After they left, I realized I was in Steve Loss's office—I was ordering him out of his own office, but I didn't care, I had to focus!

The timeframe for getting both projects completed was getting tight and there was a deadline for submitting plans which landed on a Monday. Steve and Craig decided they would work most of the weekend to complete the drawings. I said that I would be there on Sunday afternoon to help wrap it up. When I arrived, it looked like a frat house in that office! There were plans discarded and scattered across the floor; there was a pizza box with some crusts and a stale piece of leftover pepperoni pizza lying open on the one desk. They both looked a bit worn. They had spent twelve or thirteen hours there on Saturday and then came back on Sunday morning to continue! I had never worked with a team that was so devoted. They went home Sunday afternoon and I finished my checking and corrections.

We got the plan submitted and I believe the planning commission meeting that dictated the Monday deadline was the one mentioned previously where we were waiting on comments from Mr. Mean Shultz on July 3, when the National Battlefield Tower was demolished. We eventually got that plan approved, but there were so many requests for additional information, conditions, and changes to the design, that in the end, the approval was anticlimactic.

Nevertheless, Dave Davidson showed his appreciation for our efforts by taking the whole office out to lunch. As he congratulated us for a job well done, I informed him I was taking the project team, including the field surveyors, Ty and Kegan, out for some beers on the company credit card to reward them a little bit extra.

We found some bars open in downtown Gettysburg and had a few beers at each one. We ended our little soiree by going into the Hamilton Tavern on Chambersburg Street. That establishment is now a great Irish bar and restaurant, the Garryowen, but back then it was a neighborhood bar that looked like a house from the front. Some would call it a dive bar, based on not only the appearance but the following facts: they allowed smoking, the jukebox had a lot of country songs from the '60s and '70s, the clientele all knew one another and usually met there around noon each day, and when you walked in as a stranger, everyone—and I mean everyone—stared with much consternation!

By the time we got there, our group only consisted of Craig, Steve, and me, and as we walked in I could feel the eyes upon me. We were all feeling fairly jovial, so we didn't really care. As we ordered our beers I actually saw some of the bar patrons looking us up and down. In those days a necktie and sometimes a jacket were the normal uniform of the day for working in the office. We were smart enough to take our ties off, but our dress shirts and pants betrayed our efforts to fit in.

The one funny line which showed the nervousness that we were feeling was when Craig ordered a Miller Lite, and they asked draft or bottle. Craig just looked around and nervously answered, "Yeah." Everyone within earshot chuckled and the one old boy sitting nearby said, "Hell, give 'im one of each." That made us all laugh hard as Craig just flashed his characteristic wide grin.

We headed to the back room where a few folks were playing darts. After a while, they started a polite conversation with us which soon had us discussing bullfights on TV! Once they spoke directly to us, we could identify them as guys. Up to this point, I honestly couldn't tell if one or either were male or female. They weren't classic transvestites or anything, but they both had rough looks and long hair, so it was hard to tell. The one guy said, "When I was a kid growing up in Texas, we used to watch bullfights from Mexico on the TV every afternoon." This was a bit surreal, but we were drinking and just going with the flow. Eventually, they asked what we did for a living and commented on how we were, "... wearing business suits and all." I guess they didn't see anyone in there before dressed as we were, but it was definitely not a suit!

We had a few more beers and then we got to meet the famous "Goldie." I had not heard of this guy before, but apparently, he was well known among the drug user community. When he smiled, we saw his eponymous array of gold teeth. He was mostly speaking with the two other folks described before, so I didn't pay much attention to what he was saying until I heard him asking for the time of day. We told him that it was about 3:55. He said, "Oh yeah, I got time; I need to go home and get a shower, but I don't have to be at work until 4:00."

Someone pointed out that five minutes was barely enough time to get to work let alone go home and get a shower. He just shrugged and replied that sometimes he's late! As we were finishing up our beers, some people came in the back door. They looked hard at us, realizing that we didn't really belong there. One very large African-American man came in carrying a box of apples. Craig looked at him and said, "Hey, there's the box of apples that we ordered," and started to laugh. The man holding the box and everyone else just stared at Craig. They continued into the bar without saying a word and I shot a glance at Craig and said, "Let's go." We downed our beers and headed out. When we got outside, I slapped Craig in the arm and asked why he would say something like that. "Oh, that's how you make new friends," he replied and gave that wide, toothy grin of his!

Lessons Learned

We were happy to be done with that part of the approval for the subdivision plans on the Adams Commerce Center, but then there was the construction phase. Thankfully Paul Sauers was there for that and he was happy to take over on the construction plans, specifications, bidding, awarding the contract, requests for payments, and all of that.

After the whirlwind affair of the design phase, I had the unenviable task of collecting the fees from ACEDC. As stated earlier, Cathy Cresswell was the executive director ,and I had to negotiate these additional fees on a project which I had allowed the scope of work to get out of hand. I was young, inexperienced, and naïve enough to think that Cathy's word and promises of, "Just get it done—we know this is outside of your contract scope," was going to be good enough to get paid for whatever we did. There were a lot of negotiations and meetings where we reviewed emails, faxes, bills, and other documentation. It was a lot for a young project manager to work through. I would offer to go back to Dave Davidson and ask him if we could write off certain charges that Cathy did not necessarily think were valid.

Dave taught me a lot through that process and when I had to admit that I didn't properly document certain things, he would concede that we would write off half of it to appease the client, with the admonition that I shouldn't have done it that way and I should never let that kind of thing happen again. But he did it with such poise and grace that I felt like I was getting off easy. I wasn't used to that. It just wasn't Dave's style and that method worked well for me and some others who were conscientious about our work. There were, and are, others who need someone to watch over them with a big stick, but that's another book (*Managing Difficult People*—maybe I'll write that someday!)

I will also point out, Cathy was very respectful but also shrewd and fiscally responsible for her corporation. I often said that the ACC would not have happened—at least when it did—if not for Cathy Cresswell. We came to an agreement and got paid for most of the work. We wrote off a nice chunk of the fee, but we also recouped some money by doing the Pella site plans and structural design. By doing those plans, we also set ourselves up to do other projects in the commerce park, including the office for ACEDC.

Renn Kirby Chevrolet

One of the projects that we eventually landed in the Commerce Park was designing an access drive off of US Route 30 for Renn Kirby Chevrolet. During that process, we did several associated design plans for land development and erosion and sedimentation control. While working through the approval process for the access drive, which required approval from PennDOT, we also had to get approval from Straban Township. Mr. Mean had been replaced as the engineer of record, Glenn had been replaced as the township secretary and permit officer; in fact, all of the supervisors and planning commission had been replaced, but one thing remained the same: Straban Township, by and large, did not want development!

The interesting part of the story is that a woman by the name of Sharon was now the chair of the supervisors and she also sat on the planning commission. She also happened to be the sister of the former planning commission chair, Riley. She seemed to be a very nice person when we had a "workshop meeting" (meaning it was not public and no action could be taken) to discuss the details of what we were proposing. The township engineer representative, the traffic engineer representative, two supervisors, and the township secretary attended. We had a good meeting and the consensus was that if PennDOT would allow it, the township would give their approval.

When we met with PennDOT engineers in a pre-application meeting to discuss the proposal, they said we would first need a letter from Straban Township saying that this would be allowed. If this sounds like a "Catch 22," then you're paying attention and thinking the way I did at the time! My brilliant idea (hey, I thought it was brilliant, but you'll soon see that I was wrong ...) was to submit a sketch plan with a letter outlining the discussion at the meetings held prior and with a statement that requested that the township approve the plan, only regarding the access drive and explaining that the approval would not be construed as land development approval. We further stated that any grading or construction would need to be reviewed and approved by the township and would be done in accordance with their ordinances. We also asked to be put on the agenda for the next planning commission meeting.

I appeared at the next regularly-scheduled, public meeting of the planning commission to explain our proposal and see if we could get their approval. When our matter came up on the agenda, I stood and explained our proposed access drive and the need for an unofficial approval of the sketch plan. The chair of the commission looked at me and asked, "Mr. Gladhill, how long have you worked at C. S. Davidson?" I found that to be an unusual opening question, but I answered, "A little over seventeen years," to which the chair, Darren, offered, "I guess you don't do much work in Straban Township!" That got my motor running and my mouth started going faster than my mind could keep up with, but I managed to say something about doing a lot of work there, including the ACC, but nothing recently.

He said, "Well, if you did, you'd know that we don't allow any direct access to Route 30!" I should have said, "How stupid do you think I am—I know that you allowed a motel to have an access drive just across Route 30 from where we are proposing one," but I didn't. I simply said that we discussed this with members of the board of supervisors and their engineers. As I did so, I looked to Sharon to confirm what they had told us: namely that they would concede if PennDOT would approve of it. Sharon simply said, "Well, Eric, you know we have our reservations about this." When I tried to remind her of what was decided at the workshop meeting, she just shook her head and expressed her opinion that they wouldn't want that access drive there.

We were back to square one—another workshop meeting! This time, the township asked the Adams County Office of Planning and Development to research the issue and apparently bring enough ammunition to shoot the proposal down. The lady from the county office had a whole litany of ordinances and conditions from previous plans, etc., to disallow our planned driveway. I had to resort to all of the documents I had to contradict her claims—she tried to pull out an old version of our final subdivision plan for the ACC! When we discussed certain terms and conditions set forth on the recorded plan, Darren questioned why I was making a distinction between an access drive and a private driveway. I pointed out that the zoning ordinance had different definitions for those terms. He accused me of being obtuse. I had to remind him that certain words have certain meanings, that's why we use certain words in certain situations!

At the writing of this book, we are still in the approval process for this drive. The township keeps trying to prevent it. We've changed it to a right-in only and PennDOT has made some comments on the highway occupancy permit application, so we've been at this for four or five years and we just keep trying to get approvals!

Royal Farms Store

There was only one time in my career as a project manager for land development plans that our plans were denied approval—that was in Littlestown Borough. Most times, the governing authority will make it difficult to get approval by making the timeline drag out or by making conditions that make the project cost-prohibitive, but as long as the proposed development meets the ordinances, the approval is obtained. I'll preface the story about the application that was denied by telling another interesting approval in the Borough of Littlestown.

Littlestown Borough is a small community in the south-central section of Adams County. I guess you could say they're not progressive, but they do have some businesses and some good people there; it just seemed that for a long time, there were people in the government that didn't want much development to occur.

We were approached by Royal Farms, a convenience store chain based in Baltimore, about doing a land development plan for them at the north end of the Borough of Littlestown, in around 2003. I was happy to have the work for our survey crews, so I plunged in head-first as the project manager. We worked with the architectural plans under the direction of the vice president of the Royal Farms Corporation to develop a site layout.

When we first took the plans before the Borough Planning Commission, a man who was the chairman tried to find something wrong with our plans. He would look out over his thick glasses and point out something that he thought would prevent the approval of this proposal. Our old friends at the Adams County Office of Planning and Development (ACOPD) prepared a comment letter that attempted to point out major problems with our plan.

We had paid a landscape architect sub-consultant to design the required planting on the site. The planners at the ACOPD pointed out parts of the plan that they considered a deficiency regarding the landscaping ordinance. There was a section of the ordinance that required vegetative screening along the public streets on which the development bordered. Their interpretation of that section was basically that the trees and shrubs had to completely hide the store and parked cars!

I pointed out that what the ACOPD was requiring was a green tunnel on each street that had commercial development and that motorists couldn't see what type of business was just off the street where they were driving. I had seen this type of development in a planned community in Florida, known as Palm Coast. We visited there at times and found it very difficult when we would drive along the street in search of a restaurant, a grocery store, or whatever service that was needed, only to see trees and shrubs with a little sign near the entrance listing the businesses within that complex. By the time we could see the sign, we had no time to decide if we wanted to turn into the parking lot and signal our intentions to the other motorists!

After several sessions of discussions, revisions to the plans, and my arguments that we were using a registered (licensed through the

Pennsylvania Registration Law) landscape architect, and they were more qualified to design plantings than a planner from the county office, we finally got through that hurdle.

The other big point of contention was the PennDOT approval of our planned access drive onto a state highway. This sounds a lot like the Renn Kirby-Straban Township project where PennDOT agreed to approve the design, but we needed the local government to approve it as well. The borough's engineer pointed out a part of the traffic study, which tries to determine the effects of a proposed development or change to a highway, five years into the future. One of the future traffic counts showed that the cars which were stopped at the adjacent intersection, which had a traffic light, would "stack" to a point in front of our proposed access drive. This was enough for the borough officials to protest and not approve the plan.

We had to have several workshop negotiations between attorneys for both sides to prevent a "Right Turn Only" sign at the driveway. The developer would be required to conduct another traffic study at some point in time (this was a major point in the negotiations), and if the traffic count numbers dictated, the sign would have to be installed!

The ACOPD also commented on the fact that we were proposing more parking than what was required. Most zoning ordinances have requirements for parking based on the use, size of the building, or other metrics, such as the number of employees, etc.

Royal Farms had a practice of providing more than enough parking. Their purpose was to have space for contractors to meet their crews at the store in the morning before they went to work. This meant a better chance for them to buy coffee, breakfast, gasoline, and whatever they needed. The county planners countered that this was not "Best Management Practices" (BMPs), which was a buzz-word used by planners and the Pennsylvania Department of the Environment, in regards to stormwater management. The idea was that first, the designer should try to minimize the impacts of stormwater runoff by reducing the impervious coverage.

The evening of the meeting where our plan was being further dissected and eviscerated, the planning commission was reviewing some zoning matters. Two of the requests for a zoning waiver were regarding the amount of parking being provided for businesses in the borough. When our project was being discussed and I was arguing the amount of parking being provided on our plan, one of the planning commission members remarked on how stupid it was to ask us to reduce the number of parking spaces when they just had two requests to allow less parking than what was required because the applicants could not provide the parking spaces due to limitation on their property. Here we were providing more than enough and we were being told we shouldn't do that!

When we had worked through most of the objections to our proposed development, there was one more meeting with the planning commission to get their recommendation for approval; the borough council would have the final decision on whether to approve it or not. These meetings are public and are advertised as a normal practice because it is required by law. Most times, the only people to attend the meeting are the land development applicants (developers) and their representatives (engineer, surveyor, and/or attorney), but sometimes, the concerned citizens show up! This was one of those times.

The situation with this property where the store was proposed, was that it was zoned for commercial use, but there had always been a single-family house there. The property was adjacent to a housing development, which had just been built in the past ten years, and had hundreds of new residents. Most of these residents didn't realize the street that led to their homes went by this commercial property, where a high-volume convenience store would be built.

A lot of these residents came to the meeting to voice their opposition to this proposed development, but they really had no legal standing to stop the approval. Nevertheless, the law allows for them to speak and we were forced to sit silent and listen unless we were asked to interject. Fortunately, we had a representative from Royal Farms come to the meeting to explain their reasons for building the store and why it was designed the way it was. While the local citizens were speaking, I could understand their concerns because they didn't know the zoning laws, the procedure for approvals, nor the intent of the developer.

There was, however, one man who seemed a bit mentally unbalanced. He made ludicrous accusations about the proposed store, like the idea that with all of that extra parking, it would attract drug gangs who would "set up shop" in the parking lot and the whole neighborhood, and maybe even the whole borough, would be ruined! He revealed that he worked as a roofing contractor in the Baltimore area and he knew how these things worked. From the looks on people's faces in the crowd and the murmuring going on as he spoke, I could detect that he was making himself look crazier as he continued to speak!

He brought up a safety concern about the intersection at the edge of the housing development and adjacent to the proposed store. He mentioned how often the Penn State Trauma Center or Pennsylvania State Police helicopter landed near that intersection to transport victims of car accidents to the trauma center at York or Hershey Medical Center. One of the members of the planning commission was also a volunteer firefighter, so he explained that there was a helipad built in the field next to that intersection for the purpose of loading patients onto the helicopter there for transport to the hospital. The accidents didn't occur there; the ambulance brought them from the accident site to the helipad—it just happened to be next to the intersection. Everyone soon realized that this guy was just mad—perhaps in more ways than one! The plan was eventually approved and the store sits at that intersection today to provide convenient services to the people of Littlestown and the motorists who pass by.

Denied!

As prefaced earlier, there was one plan for which I was the responsible project manager that got denied, and it happened to be in Littlestown Borough. My friend, Mark Keller, had a property in the borough that was vacant, except for a small picnic pavilion, that he wished to develop into townhouses. I knew Mark from Freedom Valley Worship Center where I attended and played drums on the worship team. Mark was the guitarist, singer, and leader of the worship team. I was happy to be able to provide our professional services with the intent to make a friend's plans and dreams become a reality.

The first challenge with this property was a big drainage channel that traversed the rear portion of the property. We had our wetland sub-consultant take a look at it and determine that it wasn't protected waters. Just to be sure, he had representatives from the Army Corps of

Engineers and the Pennsylvania Department of the Environment meet there and agree with his determination. Once we knew that we could run that water through a pipe and build a parking lot over it, we felt good about making this plan happen.

Our old nemesis who chaired the planning commission at the Borough of Littlestown once again tried to find some reason to deny approval of the plan as he, once again, glared out over his thick glasses. He made several attempts at pointing out some supposed shortcoming on the plan and then grinning like the Grinch who stole Christmas. He had plenty of ammunition provided by the borough engineer and the planners from the county. Then to make this more of a challenge, the neighbors to this property complained about the idea of having townhouse apartments near their homes.

One woman even wrote a letter to the local newspaper and complained about how this was going to ruin the natural beauty in the borough. She lamented that her kids used to chase butterflies in that field—never mind the fact that they were illegally trespassing on private property! We attempted to work through the comments made by the reviewing agencies, but the one thing that the reviewing engineer wouldn't back off of was a requirement for all public streets to be at least 32 feet wide.

We were not proposing a public street; we were proposing an access drive. The problem was that the borough ordinances did not provide any definition or even mention of an access drive. Most municipalities recognize that a driveway on a private property that allows use by multiple parties to have ingress and egress is an access drive. We showed a drive that was designed to be 24 feet wide. We showed all types of design manuals which showed that it was more than adequate for two-way traffic.

Mark had a local attorney working with us, who was another friend of mine. The attorney provided good arguments against the comments provided by their engineer. The one comment that would not be withdrawn was the required width for a public street. Our attorney clearly pointed out that the section of the ordinance being quoted was not applicable, so we finally got the planning commission to give their recommendation for approval.

When we went before the borough council, we were ready

to accept some conditions to the approval, but instead, one of the councilmen read from a prepared statement as his motion to deny the plan! I was so angry that all I could think of was jumping across the table and choking this guy! His remarks and his stance on the subject were so misguided and completely wrong that I felt my blood pressure rise as I calculated if I could dive across the table, hit him in the shoulders with all of my weight, and drive him into the wall. The motion passed on a 3-2 vote. Our attorney asked some questions with a clenched jaw as his face turned a deep shade of red! As I said, he is a friend of mine and we had worked together on several legal matters; I had never seen him get mad at all, now here he was, ready to spit nails. He declared in no uncertain terms that their decision was wrong and that a court of law would certainly overturn it.

We didn't know if Mark would want to file a lawsuit in order to get his approval or not, but after some conference, Mark instructed the attorney to move forward with it. When the judge brought the two parties together to discuss the matter, he looked at the borough solicitor and said, "You know these matters rarely go in favor of the municipality." He then instructed the solicitor to go back to his clients and have them overturn their decision. This was done and we finally had our approval. That was eleven years ago and it was never built. Recently someone was looking to buy that land and build the townhouses. I guess we'll see if that occurs or not.

Fight for your Right

I never really thought my career would become one fight after another with municipalities; it didn't really, but there were months on end where it felt like that. The truth is, while I was staying busy and trying to keep others in our firm busy by working for private land developers (even when those developers were just regular folks and sometimes a utility company), I was also serving as the representative of C. S. Davidson, Inc. as the engineer of record for the municipality for three different clients. So, at various times, I would sit on the side of the table with the elected or appointed officials for a municipality and represent their interests, and then the next day, I might be asked to attend another meeting in another jurisdiction and represent the developer and their right to use their land. In all honesty, I will always feel that the property owner should be able to invest money and use their property in any way they want within the confines of the law. I also think that the laws should not be too restrictive but should protect the rights of all citizens. Too many of the laws created by the municipalities were created with good intentions, but when they are applied to normal land-use proposals they make no sense. One case in point follows.

Columbia Gas became a major client for C. S. Davidson, and I had the privilege of managing this great client. On one occasion they needed to expand one of their regulator sites. These are the fenced areas with some pipes and fittings sticking out of the ground. The purpose of the above-ground appurtenances is to regulate the pressure of the gas being distributed to the customers. The subject site was located in Springettsbury Township, York County, Pennsylvania. It was also the site of a roadway widening and realignment. The regulator was located within an easement on an unusual piece of ground.

The property was part of a large farm, but the new roadway had bisected a corner of the property and had also encumbered the remaining corner of the property (about one half of an acre) with a stormwater easement that took up most of the land that wasn't already used by the gas regulator site easement. Columbia Gas regularly asked us to serve as land agents in acquiring rights of way on their behalf. Since they needed to expand their regulator site, they asked us to acquire an area of easement adjoining their existing site. The widow who owned the land had no use for the remaining land that lay on the other side of the new roadway and PennDOT right of way that was bought in fee simple.

When I contacted her by mail, she had her attorney respond and inform us that his client would transfer her interest in the property to Columbia Gas for free. This seemed like a great deal! I surmised that since the piece of land had boundaries already created by the new road and the existing boundaries of the old deed which ran along the old roadways, I would just write a description based on our survey, include a plan showing the existing features and ask the attorney to prepare a deed. The attorney, John, said that I needed to get some kind of approval from the township or we could be guilty of an illegal subdivision.

I sent my suggested legal description, drawing, and letter of explanation to the township planner, and I also sent the same information to the tax mapping office asking them to assign a parcel number so that the deed of transfer could be written. Within a few weeks, I received a reply from the tax office with a duly assigned parcel number. I waited a few more weeks and got no reply from the township. I notified John and suggested that we were ready to move forward with the land transfer. He said he would not do so unless he received some approval in writing from the township.

I contacted the township and they asked me to attend an in-house meeting with the township planner, engineer of record, and public works director. This seemed like a good opportunity for me to explain the unusual situation and obtain their written approval. They wouldn't really be approving anything, just acknowledging that due to the right of way taking by PennDOT, a parcel of land had been created and there was nothing they could do to reverse it, change it or deny that it existed.

After much discussion, the township engineer, who was appointed and served as the engineer of record, just as C. S. Davidson did as part of our professional practice, suggested we would still need to submit a "simple subdivision plan." This was a term used in their ordinance for the creation of two lots which front on a public road, but I replied, "There is no such thing as a simple subdivision these days." They all chuckled and said that it would be necessary to go through with a final plan (which has a lot of detailed requirements) which would need to be reviewed by the township engineer, York County Office of Planning, and the township planning commission.

Of course, this whole process also involves public meetings before the planning commission and the township board of supervisors. Our client authorized us to move forward with the preparation of the plan and pay the application and review fees involved. This seemed ludicrous and completely unnecessary to me, but we had no choice but to move forward. The township engineer—the one who claimed that the final plan was necessary—had plenty of comments on our drawing. Some of them were just nit-picking comments regarding minor points of content, but the one was extremely onerous; he claimed that we needed to provide a complete boundary survey of the remaining farm which consisted of about one hundred acres! We made our arguments in writing, by phone call, and in-person until we had no recourse but to appear at the meeting of the planning commission and allow them to decide on whether this additional survey work was really needed.

The meeting room at Springettsbury Township has the appearance of a courtroom or congressional hearing room. It has a raised platform at the front of the room with a polished wood counter, similar to a judge's bench, and wood paneling on the walls behind it. The planning commission members each have microphones in front of them and the person appearing before them is required to stand in the front and center of the room with a microphone on a floor stand.

I have to say that speaking before a panel in a public meeting can be intimidating but sometimes it's more comparable to a youth baseball league meeting being held at the fire hall with a folding table, folding chairs, and some good ole boys sitting there in blue jeans behind the table—very informal—but this was more than intimidating. I don't think I felt any more nervous when I appeared as an expert witness in federal court.

Early on the agenda was another plan by our firm for an expansion of York Central High School, which was being represented by one of our young engineers. He explained the need for a building expansion and additional parking even though the school was just built in the last year. He was being chided for bringing this plan for approval when this should have been included in the initial proposal. All I could think was, *These folks are tough.*

When it was my turn on the agenda, I explained that the plan before them had no proposed features; every line and every symbol represented existing conditions. They still asked a lot of questions and eventually approved our request for a waiver from the section of the law that required a survey of the parent tract of land from which the new lot was being created. Of course, it was not applicable because we weren't creating anything new! Just another example of how the law can create hardships or actually how a township engineer making a misapplication of the law can create problems.

Another Difficult Engineer

There was another occasion when a township engineer that we'll call, Knothead, tried to make extra work for me and Columbia Gas.

There was an extension of a gas main being planned, and while we were doing the design and planning for this, Columbia Gas decided they would like to move their regulator facilities away from the gas line transmission station. When we met with Mr. Knothead to discuss the need for grading permits, stormwater management, etc., he claimed that this may be considered land development.

Land Development is a legal term that is defined in the Municipalities Planning Code. I won't pretend to be a lawyer and I won't quote the section of state law here, but I can assure you that there is nothing in the law that can be interpreted that facilities for a public utility constitute land development. When I told Mr. Knothead that, he replied that it would be up to the township board of supervisors and their solicitor to determine. Even when I paraphrased what the legal definition was, he just countered that we weren't going to argue that point. I agreed and said that we had no reason to argue it, the law was clear.

Of course, I wasn't there when he discussed it with the township board, but they sent a letter saying that land development was required. The attorney for Columbia Gas had to provide a letter with several case histories listed which refuted this requirement. Eventually, the township solicitor acquiesced and we acquired the necessary grading and stormwater permits. This type of behavior has caused plenty of heartburn for me and my clients over the years.

Surveying Under Fire

Every surveyor has a story of being shot at in the line of their duties. Some of the stories sound fishy, like maybe someone was shooting and an errant shot came near to the surveyor, or maybe the property owner was just shooting over their heads to scare them, but I only know a few surveyors who were actually the target with the intent to injure or kill.

One day in the summer of 2012, I was investigating some boundary evidence that our field crews had surveyed on a farm that was about to be bought by a private foundation to be turned over to the National Park Service, Gettysburg National Military Park. The one boundary adjoined a campground where they stabled and pastured horses that were used to give tours of the battlefield. There was one corner, in particular, that was not found, but the crews had located a large tree with old barbed wire embedded into its trunk. As I walked along the line adjacent to the campground, I stepped into the tree line to look for other evidence of old wire grown into the trees.

I heard gunshots that sounded like they were in the direction of the corner where I intended to go. It was pretty clear that the shots emanated from the campground property and not the farm being surveyed. The sounds of the gunfire seemed to be about two hundred yards in front of me (I always tend to think of distances in yards when it comes to shooting guns, because target shooting is usually expressed in yards and not feet), so I wasn't too concerned. It was summer so I assumed someone was killing a varmint; either groundhog or snake.

The shooting continued as I covered another one hundred yards and I began to think that someone was shooting targets, but I couldn't understand why they would be doing that on the campground property where families were staying in campers and tents. I suddenly heard what sounded like a bullet hitting some branches in the tree line which was about twenty-five feet to my right. As I went on guard and started looking through the trees to see who was shooting and where this was coming from, another shot rang out as I heard a bullet hit some brush in front of me and whistle past my ear as only a tumbling bullet can sound. I had heard this while hunting, but never this close! I hit the dirt and yelled that there were people "up here." There were a few more shots and I hollered as loud as I could in the direction of the shooter.

As I lay on the ground, thinking how close that bullet was to my head, I tried to calm my breathing and settle my nerves. I reflected for a moment what it must have been like for the soldiers who fought on this ground during the Battle of Gettysburg when they didn't have any choice but to march forward.

When the shooting stopped, I walked quickly toward my vehicle and drove to the campground store and office. I told the ladies who were working there of my recent experience to alert them of a shooter on their property. They told me the owner lived back in that corner of the property and he had built a shooting range where the local police were invited to come and practice with their handguns. They called him on the radio and told him of my report, he replied that he would be there in a few minutes.

When the owner arrived on a golf cart, I could see that he dressed like a western sheriff. He introduced himself and allowed me to do the same. When I told him of my duties in surveying the property next to his and reminded him that we had mailed a letter to inform them of the survey, he thought for a minute and said, "Yeah, I got your letter, but you said you would call me to let me know when you were coming out." I said, "No, the letter informed you of the survey that would be performed and it stated that you could call us if you had any questions or wished to meet our crew on the site."

He then told me that he was a retired sheriff and had constructed a shooting range for the local police to use. He assured me that the bullet that I had heard (he was very doubting of the facts of my report; not even accepting the fact that I had heard a bullet) did not come from his shooting range. He told me about how the backstop was all dirt and it was fifteen feet high, etc. I asked if someone had been shooting there in the past thirty minutes and he confirmed that there was someone back there shooting. Then he offered to show me his range.

I looked at the backstop and it was built for safety, just as he said. I don't know how the bullets got to where I was, but my conclusion was that the shooter (who had left by now) fired in the wrong direction and not directly into the backstop.

The property owner allowed me to walk through his patch of woods into the corner that I wanted to investigate. There were other targets that I saw back in the woods and felt that the shooter was probably off of the range and shooting at random targets. We left with a disagreement about what had happened, but I was just happy that my life was spared and I didn't get shot!

Epilogue

Now that I've been at C. S. Davidson, Inc. for more than twenty years, and have written many of the funny stories that I can remember from my long career in surveying, one thing that I learned after all of these years is that not every one of your co-workers is your friend. Some managers that I've worked with think we should be one big happy family and we should all be great friends that all get along together. Ty Martin taught me several good lessons about life in general and work relationships, in particular. One thing he taught me is that you can work with someone and be a teammate, but you don't have to like that person. I've come to relate to my co-workers as teammates; some of them are friends and some are just people who I work with.

I like the analogy of a firm being like a team; we are all working to "win" the contests in which we are placed. Some teams have members who will fight each other and yet stand by each other in the heat of the game to do what is needed to win. If a quarterback is upset with a wide receiver because of some ill-chosen words or differences of opinions, he will still throw a pass to him if he's open. Teammates work well together whether they are friends or not.

One such example that I like to recall is the Oakland Athletics in 1974. They were on the verge of winning their third World Series in a row when a fight broke out between two of the pitchers (it doesn't really matter who they were, but if you're interested, it was Rollie Fingers and John Odom) before one of the World Series games. There were punches thrown and landed, Rollie got knocked into a locker and had a gash cut into his head that required five stitches to close it! In the game, they all played hard together to help each other win. They beat the Dodgers in five games.

As I went from one job to another over the years, people used to ask me why I kept leaving to find another job. I told them that I would work hard for a company as long as they appreciated me; when I felt unappreciated, I went somewhere I would be appreciated. I learned along the way, that no matter how valuable a person might be, they can always be replaced. When someone finds another job or dies, the company will figure out a way to fill that void and the projects will go on. My departure from each of these positions was nothing more than a minor blip on the financial heartbeat of those firms.

Now, I can reflect on some of the good times and bad times. Some of the great people in this book have gone on to their final reward, including, my son, Dusty.

Earl Clouser, Ed Warfel, Ed Pinto, Bill Peck, Bill Fox, Paul Sauers, Craig Angle, and Randy Snyder have all passed away from this life. I still have fond memories of them, and these stories may not paint all of them in the kindest or most flattering light, but I simply wanted to tell the stories that made me laugh, shaped my character, and gave others some edification.

Ty Martin and Greg Myers turned out to be some of the best friends ever. They were the ones who called me on the day that I lost my son, and they continue to be the ones who support me because they knew Dusty well and can share my sorrow.

I now play drums in the band Jalopy Deluxe and find great friendship with the other two members of the band, Bill Bruder and Kelly Gray. I've also made some good friends through my involvement with Rotary.

I sometimes wonder if surveying was really the correct vocation for me, but this is what I've done and I can be happy with what I've accomplished without a college degree. I hope that the reader will realize I have done my best to live a life of service to God, my family, and the world in which I live. And because of all these situations and people ... I found my boundaries.

About the Author

Eric Gladhill, PLS, serves as a Senior Client Manager at C. S. Davidson, Inc., in their Gettysburg office. He is licensed as a Property Line Surveyor in Maryland and a Professional Land Surveyor in Pennsylvania. He serves as a Director for the Pennsylvania Society of Land Surveyors (PSLS) and is a member of the Maryland Society of Surveyors (MSS).

He has been married to the same woman for 37 years, his faithful wife, Trudy. They live in the mountains outside of Gettysburg, Pennsylvania.

Mr. Gladhill has over 41 years of experience in the surveying profession and has become a writer as his career has taken him into the role of project manager and client manager

www.ingramcontent.com/pod-product-compliance
Ingram Content Group UK Ltd.
Pitfield, Milton Keynes, MK11 3LW, UK
UKHW021905190726
13853UKWH00002B/529